Grades 5–8

MATH AMAZEMENTS

Astounding Investigations Uncover Math in Your World

Pamela Marx

DEDICATION ....

This book is dedicated to Mark
and to Siv Si, whose teaching energy inspired it.

ACKNOWLEDGMENTS ....

Grateful thanks for help in the creation of this book go to Martha Gustafson, Lisa Skylar, Jessica Skylar, Katherine Skylar, Melina Wyatt, Britta Gustafson, Bobbie Dempsey, Megan Goldstein, Holly Goldstein, Ermuelito Navarro, Susan Sides, Nick Martin, Mark Goldstein, Rhonda Heth, Siv Si, Debbie Vodhanel, Fumiko Makishima, Nancy Scher, and all the other friends, teachers, and students who provided ideas, tried out activities, and gave suggestions and comments.

Good Year Books
Are available for most basic curriculum subjects plus many enrichment areas. For more Good Year Books, contact your local bookseller or educational dealer. For a complete catalog with information about other Good Year Books, please contact:

Good Year Books
P.O. Box 91858
Tucson, AZ 85752-1858
www.goodyearbooks.com

Cover design and illustrations: David Fischer
Interior design: Dan Miedaner

ISBN-10: 1-59647-071-2
ISBN-13: 978-1-59647-071-2

1 2 3 4 5 6 7 8 9 10 – ML – 09 08 07 06

CONTENTS ....

INTRODUCTION ....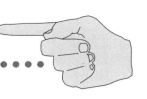

Sometimes we think that math is multiplication and division, addition and subtraction. It is equations and probability and ratios and fractions. As we get older, it is algebra and trigonometry. Math is all of those things, but it is so much more. And most of what makes up math is just plain fun and interesting. This book takes a walk through history, our environment, math thinking, and logic to provide explorations and activities that show how much math there is in the everyday things around us and how much fun it can be to search for the math in things.

The activities in this book offer great ways to stimulate your math thinking. For most people, while some kinds of math thinking and activities are easy, others are difficult. This is true for almost everybody. It simply depends upon how your brain works. Sometimes what is hard or easy depends on when you were born in the history of the world. Nobody would say the ancient Egyptians were not a smart, advanced people, but guess what? They didn't know how to multiply like we do because they didn't really need to and the method wasn't developed yet. Some of the activities in this book will help you see how math thinking has developed through the history of all people. Today, we know a lot of things by third grade that some people in history never knew, and, yet, they built the great pyramids.

Keep all of this in mind as you explore the activities in this book. You might think some of the activities in this book are really easy. Some of them will be easy for you to solve, while others, such as certain kinds of puzzles, may be hard for you but easy for your friend. This is not a book of right and wrong answers. It's a book to give you ways to think about numbers and shapes and surfaces in a new way. Above all, it's a book to have fun with while you are learning about new ideas, relationships, and concepts.

HOW TO USE THIS BOOK

This book includes thirty-five sets of activities, an answer key, a glossary, and a bibliography. The activities provide explorations in areas of math thinking such as geometry, topology, interesting properties of numbers, logic, and probability. The activities will get you thinking about geometry, symmetry, topology, math history, number properties, probability, ratios, puzzles, and games.

Each activity is described on two (and sometimes more) pages. The first page provides some background and introduction to the particular activity. The pages that follow give you ideas and opportunities for exploring the subject. In the "Your Turn" section, you will have a chance to investigate the math subject or idea introduced on the first page. In the "And Another Thing" section, you will find additional ways to explore the subject or ideas for doing different, but related, math investigations.

The book's additional features are an answer key, glossary, and bibliography. The answer key offers answers for many of the questions and puzzles. Some questions simply challenge your thinking. They may have no right or wrong answers or more than one possible answer. Others may have different answers depending upon the approach you take. A quick answer is not the point; the thinking itself is the point. Those questions in the text preceded by a number in a circle ① are answered in the answer key. The glossary offers a way to check your understanding of a term in an activity. The bibliography includes many books that provide wonderful math ideas and curiosities that students of all ages will enjoy. That's because math can be just plain fun and interesting when we stop thinking about it as a test to be survived. The Internet is also a great source of entertaining and stimulating math inquiries.

Flip through this book. Stop to read when something interests you. Pull out a piece of graph paper and see what wonders will unfold for you. Above all, enjoy!

A NOTE TO TEACHERS AND PARENTS........
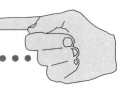

The activities in this book vary greatly in complexity. Some, such as "How the Egyptians Multiplied," are mathematically quite simple but are interesting to help young people understand the development of math thinking through the ages. Others, such as "The Four-color Map Problem," are very complicated mathematically, but young people can enjoy exploring the concept or "problem" and some of the initial thinking that goes into solving complicated math problems. Still others, such as the tangram activity, are easy for some and difficult for others, regardless of age. It's a matter of how our individual brains work. Playing with these puzzles, whether initially easy or difficult, builds abilities and skills.

The activities in this book are written so that they can be enjoyed by young people or presented by an adult who introduces the activity. They are organized so that the activities build in complexity as you progress through the book; however, everyone's brain works differently and some activities that are easy for one person may be more challenging to another. Each activity is presented on two (or more) pages. A teacher or adult can use the material on the first page of any activity to help present the activity and then provide the young person with the following pages to do an actual hands-on exploration. Alternatively, provide the young person all the pages of an activity to read and investigate. If an activity sparks a particular interest, encourage further exploration. There are many excellent math activity Web sites available, which include particular topics, such as those in this book. You can find these by using keywords in one of the familiar search engines.

Another wonderful use for the activities in this book is as a jumping-off place for group, family, or even community explorations. Teachers might use the activities in this book for special projects, extra-credit assignments and student presentations. Families might use the activities for family fun nights, youth group activities, home schooling, or even parties. Nothing gets people involved in an activity more quickly than exercising their brain cells over something they think should be easy, only to find that it isn't as easy as they thought.

Remember that the best use of these activities is to provide a stimulating and entertaining way to foster mathematical thinking and to create and develop a pleasure in the process of thinking, creating, and puzzling things out. When it's all said and done, we learn more that we remember longer when we're having fun at the same time.

1 FACTORING, GEOMETRIC SHAPES, AND PRIME NUMBERS

The Greeks thought of numbers in groups and gave them geometric names, such as *triangle numbers*, *square numbers*, and *pentagonal numbers*. Positive whole numbers can be analyzed in a geometric way all their own to find out whether or not they are prime and to find out the factors of a number.

What is a *prime number*? It is a whole number greater than 1 that has only itself and 1 as factors. Prime numbers include 1, 2, 3, 5, and 7.

What is a *factor*? Factors are numbers that can be divided into a whole number with no remainder. For example, the factors for the number 8 are 1, 2, 4, and 8: 1 x 8 = 8 and 2 x 4 = 8. The number 8 has four factors, so it is *not* a prime number.

To find out if a number is prime or not, when you look at it, ask how many kinds of rectangles you can make from it. When you figure this out, you will know whether the number is prime and what its factors are.

For example, look at the numbers 5 and 12. How do you find out if either is a prime number? First, draw a rectangle made of squares for the number 5:

Now look at rectangles for the number 12:

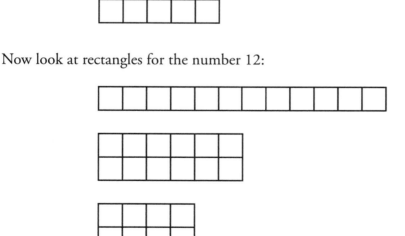

1. What are the factors for each number?

2. Is either number prime?

1

YOUR TURN ....

Analyze a non-prime number of your choice in terms of the kinds of rectangles it makes. When you have drawn your information here, turn your factoring information into a chart. It is easy to see how factoring relates to multiplication when you chart a non-prime number in this way.

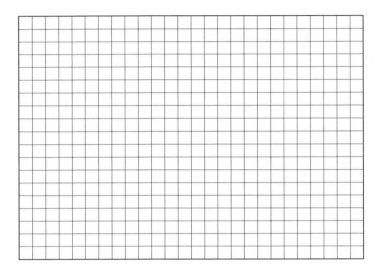

After you have done your factor rectangle analysis, turn your number into its prime factorization—that is, write your number as a product of its primes only. For example, the factors of the number 12 are 1, 12, 2, 6, 3, and 4. Forget about 1 because anything times 1 is the number itself. How would you show 12 using only its primes? You would write it as 2 x 2 x 3. You would write 24 as 2 x 2 x 2 x 3.

AND ANOTHER THING ....

Prime numbers have some interesting characteristics. For example, 2 is the only even prime number, because all other even numbers have 2 as a factor. Also, 5 is the only prime number that ends in 5 because all other numbers that end in 5 can be divided evenly by 5. Here's another interesting characteristic: If you add 1 or subtract 1 from any prime number over 3, the resulting sum or remainder will be divisible by 6. Try it.

Take the prime number 19. If you add 1, you get 20, which is not divisible by 6. But if you subtract one, you get 18, which is divisible by 6. Can you prove this rule wrong?

2 THE SHAPE OF NUMBERS— IT'S GREEK TO ME

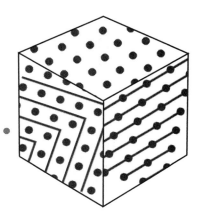

Triangle numbers? Square numbers? What in the world do these terms mean? Ancient Greek math geniuses found it fun to play with numbers and their relationships. They were mostly interested in geometry, the math of the shape of things. That's why when they looked at how numbers related to each other, they toyed with arranging them in shapes. Can you guess what they found?

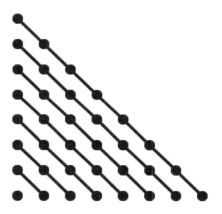

To begin, look at the very first dot in the lower left corner. Think of the first dot as number 1, the first triangle. Now, look at the two dots in the next diagonal row to the right. These two dots and the first dot make a triangle: 1 + 2 = 3. The next triangle has 6 dots—3 from the prior triangle plus 3 dots on the line of dots you add: 3 + 3 = 6.

The number of dots in each succeeding triangle is 10, 15, 21, 28, and 36. There is a pattern here. Can you figure out what the pattern is?

Now you can see why the Greeks named certain numbers *triangle numbers:* These numbers can be shown as dots in a triangular pattern. How might this relate to their reasoning for naming certain numbers *square numbers?*

Both the smallest triangle and the smallest square numbers are 1. We saw that there was a pattern to how the triangle numbers grew. The first six square numbers are 1, 4, 9, 16, 25, and 36. What is the pattern to making these numbers?

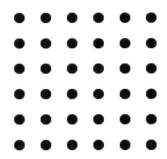

1. Can you think of a way in which triangle numbers relate to square numbers? What happens when you add any triangle number to the next higher triangle number?

$$1 + 3 = ?$$
$$10 + 15 = ?$$

2. When you look at numbers, you can often see ways in which one group of numbers relates to another. You have found a way in which triangle numbers relate to square numbers. Can you think of a way in which odd numbers relate to square numbers? Perhaps this drawing will help.

$$1 =$$
$$1 + 3 =$$
$$1 + 3 + 5 =$$

AND ANOTHER THING ....

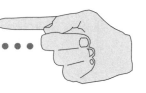

Square numbers have many cousins. You now know how square numbers relate to both odd and triangle numbers. Let's figure out how they relate to square roots and cubic numbers.

The square root of any number is another number that, when multiplied by itself, equals the original number—for example, 3 is the square root of 9 because 3 x 3 or 3 squared equals 9. It is pretty clear how square numbers relate to square roots. Look at the square numbers on the partial multiplication table below to see how square numbers relate to square roots.

1	2	3	4	5	6
2	4	6	8	10	12
3	6	9	12	15	18
4	8	12	16	20	24
5	10	15	20	25	30
6	12	18	24	30	36

(3) Now, what is a cubic number? The ancient Greeks thought about cubic numbers, too. Look at the pictures below. How do cubic numbers relate to square numbers and square roots?

The cubic number 2

2 x 2 x 2 = ?

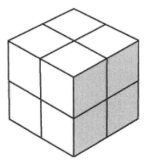

The cubic number 9

3 x 3 x 3 = ?

3 ERATOSTHENES' PRIME NUMBER SIEVE

As we learn more about math, we find out that numbers are not simply whole numbers, odd and even numbers, or even positive and negative numbers. They can be many things.

> **There are real numbers.**
>
> **There are numbers that are integers.**
>
> **There are rational and irrational numbers.**
>
> **There are prime numbers.**

The list seems almost endless. Have you heard about some of these kinds of numbers? Let's find out more about the last ones—the prime numbers.

Eratosthenes was a Greek mathematician and philosopher who lived from around 276 B.C. to 194 B.C. He had a fascination with prime numbers. What is a prime number? A prime number is a whole number greater than 1 that has only itself and 1 as factors. A factor is a number that can be divided into a whole number with no remainder. For example, 2 is a factor of 4, because 2 can be divided evenly into 4 with no remainder.

Eratosthenes figured out a method of finding all the prime numbers up to a given number such as 25 or 100 or more. It is called the *prime number sieve*. Here's how you make your own prime number sieve:

▶ First, write out your numbers in a neat way, such as 10 rows of 10 numbers to 100 or 5 rows of 5 numbers to 25.

▶ Next, cross out 1 because it is not a prime number. Why not? Because the mathematicians say so is one answer. What answer can you think of?

▶ Now, circle 2 because it has no other factors but itself and 1. After the number 2, cross out every number that 2 goes into evenly.

▶ Look at 3. Is it a prime number? Yes. Circle it and cross out every number into which you can divide 3 without a remainder.

▶ Continue this process until all the numbers are either circled or crossed out. The circled numbers are your prime numbers. What number do you circle next?

YOUR TURN

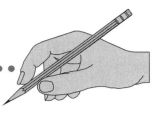

Here is how the Eratosthenes' sieve works on numbers below 25.

1̶ ② ③ 4̶ ⑤
6̶ ⑦ 8̶ 9̶ 1̶0̶
⑪ 1̶2̶ ⑬ 1̶4̶ 1̶5̶
1̶6̶ ⑰ 1̶8̶ ⑲ 2̶0̶
2̶1̶ 2̶2̶ ㉓ 2̶4̶ 2̶5̶

How many prime numbers are there in the counting numbers below 25? Now try to work Eratosthenes' sieve on numbers 1 through 100.

1	2	3	4	5	6	7	8	9	10
11	12	13	14	15	16	17	18	19	20
21	22	23	24	25	26	27	28	29	30
31	32	33	34	35	36	37	38	39	40
41	42	43	44	45	46	47	48	49	50
51	52	53	54	55	56	57	58	59	60
61	62	63	64	65	66	67	68	69	70
71	72	73	74	75	76	77	78	79	80
81	82	83	84	85	86	87	88	89	90
91	92	93	94	95	96	97	98	99	100

(1) List all the prime numbers you found between 1 and 100.

AND ANOTHER THING

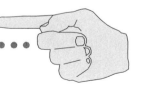

Make art by the numbers. Using the number layout above for numbers 1 to 100, investigate the patterns you make when you put different-colored dots on all even numbers, all multiples of 3, all multiples of 4, and/or all multiples of 5.

4 A HEXAGON HERE, A HEXAGON THERE

There's just no getting away from mathematics. It's all around us—just step outside and take a look. Number sequences can be found in the way plants grow. Look at their petals, leaves, and seed structures. (For activities on finding Fibonacci numbers in nature, see "A Fibonacci Search," page 109.) Geometric shapes abound in the spirals inside univalve seashells, the elliptical orbits of comets, planets, and electrons; the spherical shape of the Earth, and the hexagons of a beehive. Oops, stop right there. Hexagons are everywhere.

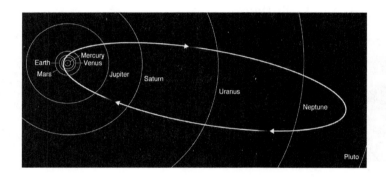

Why, you may ask, are there so many hexagonal shapes and designs around us? A hexagon is a six-sided polygon. Like the square and the equilateral triangle, each of its sides is the same length and it can fit right next to itself over and over again leaving no gaps or spaces.

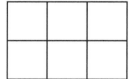

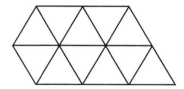

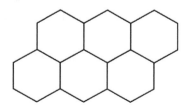

Interestingly, though, when you compare a square, an equilateral triangle, and a hexagon, all of which have the same area, it is the hexagon that takes the smallest perimeter to enclose that area. That makes the hexagon a very practical shape in nature. Insects that create honeycomb-shaped living structures, such as wasps and bees, can use less building material and energy to make their homes using this shape. Snowflakes are also hexagonal in design. Although it is said that no two snowflakes are alike, they all have six points. Might there be a structural reason that nature decided on the six points?

YOUR TURN  . . . .

Searching for certain shapes in your environment can bring surprising and interesting results. First, take a look at the natural world around you. Look at plants for hexagonal leaf, seed, and petal patterns. Keep track of your findings. Hexagons are not the only shapes honored in nature. There are many pentagonal structures and designs as well as designs in threes and fours.

To broaden your shape search, look at specimens and pictures of sea creatures. What do you find? Look at the patterns found in land and air creatures as well.

OBJECT	TRIANGULAR	QUADRILATERAL	PENTAGONAL	HEXAGONAL

What do your results show? You might choose to graph your results.

Now survey the human-made structures around you. What do you see? Do hexagonal patterns dominate or does some other geometric design appear to be more common? Graph your survey results. Does your survey of human-made objects bring different results from those seen in nature?

AND ANOTHER THING 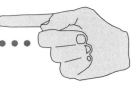 . . . .

Let your hair down and blow some bubbles. Better yet, whip up some foam with bubble solution. Press the foam under a magnifying glass or other firm, clear substance. How do the bubbles join at the corners?

(1) When joined in the foam, what polygonal shape do bubbles take on?

5 BONING UP ON RATIOS

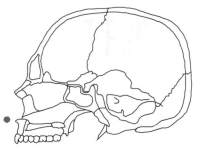

The bones in our bodies offer an interesting subject for exploring the relationship between things. Specifically, we can use bones to look at ratios or the relationship between two like numbers. You can write ratios with a colon, as in 1:2; as a fraction, as in $\frac{1}{2}$; or as a division problem, as in 1 divided by 2.

Ratios are a way of comparing two things. For example, how would you describe the relationship between the number of yolks in a dozen eggs to the number of whites in a dozen eggs?

How many yolks are there? 12

How many whites are there? 12

You might say then that for every 12 yolks, there are 12 whites. Said more simply, there is a 1 to 1 or 1:1 ratio of yolks to whites in a dozen eggs. This is easy to see if we write the numbers as fractions.

$$\frac{12 \text{ yolks}}{12 \text{ whites}} \qquad \text{reduces to} \qquad \frac{1}{1} \qquad \text{or} \qquad 1 \text{ to } 1 \quad \text{or} \qquad 1:1$$

Let's use the bones in our bodies to play the ratio game. How many bones do we have and where are they? Here's a description of many of the 206 bones in the adult body. Babies are born with nearly 350 bones, many of which fuse as we grow.

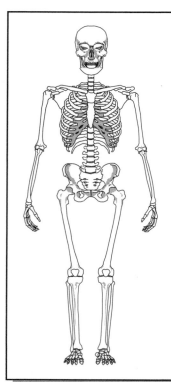

IN THE SKULL

2 nasal
1 lacrimal
1 occipital
1 sphenoid
2 temporal
1 frontal
1 maxilla
1 mandible
2 zygomatic

IN THE EAR, EACH

3 bones

IN THE FOOT, EACH

7 ankle
5 instep
14 phalanges (toes)

IN THE HAND, EACH

8 wrist
5 palm
14 phalanges (fingers)

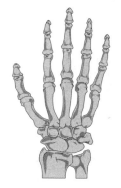

RIBS (12 PAIRS)

24 total

IN THE BACK AND NECK

4 coccyx
7 cervical
12 thoracic
5 lumbar
5 sacrum

SHOULDER/GIRDLE

2 clavicles
2 scapula

IN THE LEG, EACH

1 upper, 2 lower

IN THE ARM, EACH

1 upper, 2 lower

YOUR TURN

Use the numbers of bones listed on page 10 to come up with answers to these ratio questions.

1. What is the ratio of upper arm bones to lower arm bones? _____

2. What is the ratio of finger phalanges to toe phalanges? (Look at the egg example on the prior page. How do you answer this question using the smallest numbers possible?) _____

3. What is the ratio of ankle bones to toe bones? _____

4. If you compare the number of nasal bones (2) to the total number of skull bones (12), you get a ratio of 2:12. Can you write this ratio with smaller numbers to create another equal ratio? _____

A proportion is a statement that shows equality between two ratios. 2:12 is the same as 1:6. To write a proportion, you use double colons like this: 2:12 :: 1:6.

Let's find some more bone ratios.

5. What is the ratio of ankle and instep bones to toe bones? _____

6. What is the ratio of lumbar vertebrae to sacrum vertebrae? _____

7. A baby is born with about 350 bones some of which fuse as the baby grows. An adult has 206 bones. What is the ratio of adult bones to bones in a baby? _____

8. If you round the number of baby bones to the nearest fifty, what is the ratio of adult bones to bones in a baby? _____

9. In making a comparison between adult and baby bones, what is the most useful ratio for making a quick comparison of the difference? _____

Make up three ratio problems of your own.

AND ANOTHER THING

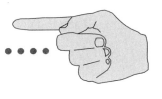

Make up your own "bone" math problems for friends to solve. Starting with the skull, make up two problems each about skull bones, arm and leg bones, back bones, hand and feet bones, and rib and ear bones. Your problems can compare bone groups. Write some ratio, averaging, division, and multiplication problems. Try to write some problems with an unknown variable to solve.

6 IS SYMMETRY JUST SYMMETRY?

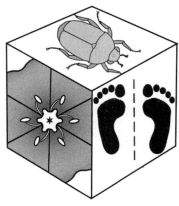

By the time you were in second grade, you had learned what symmetry is. It's found in the shape of an insect—that is, if you draw an imaginary vertical line down the middle of a bug, the two sides will mirror each other.

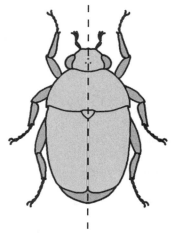

This mirror-image symmetry is called *reflective symmetry*. If you put a mirror on that line down the middle of the bug, the whole creature will appear and both sides of the bug will look exactly the same.

Another kind of symmetry is *rotational symmetry*. You can find this type of symmetry in a figure by rotating it or turning it on an axis. If, for example, at 60, 90, or 180 degrees, the first shape matches the second shape, the design has rotational symmetry. Look at the picture on the left. Rotate the picture 90 degrees and the picture matches up with itself. It has rotational symmetry. Is the same true for the diamond and the flower? How many degrees do you need to rotate them to find out?

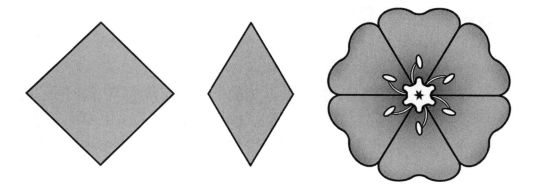

We find still another kind of symmetry in the footsteps we take. If you hop on both feet, as in the diagram on the left, you can move the prints on either side of the dotted line up or down and they retain their symmetry. Clearly, the feet have reflective symmetry, but they have another kind of symmetry as well. When you move the feet up or down on either side of the line, they match up again. So this picture also demonstrates something called *translational symmetry*.

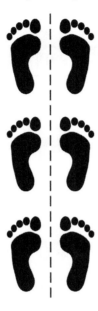

If you walk normally on both feet, your footprints are not reflective as they were when you hopped, but they are glide-reflective—that is, you can move the prints up or down in a vertical line and find a reflective position.

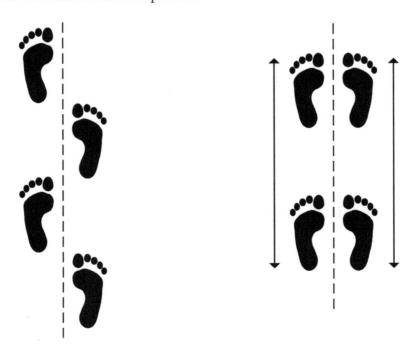

YOUR TURN ....

Train your eye to find the different types of symmetrical images around you. Do a symmetry survey of your environment.

REFLECTIVE	ROTATIONAL
TRANSLATIONAL	GLIDE-REFLECTIVE

AND ANOTHER THING ....

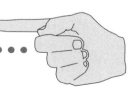

Create three rotational symmetrical designs using first an equilateral triangle as your starting design, then a square, and finally a regular hexagon.

(1) How many degrees must you rotate each design to find a symmetrical position? Explore rotational symmetry using pentagons, octagons or other polygonal shapes.

7 THE MAGIC LINE— A MATTER OF SYMMETRY?

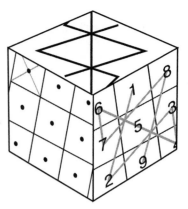

A magic square is a group of numbers arranged in a square so that the sum of any row, column, or diagonal is the same. (See "Magic Square in the Making," page 69.) In the early 1900s, architect Claude Bragdon found that by connecting the numbers in a magic square consecutively beginning with the smallest number, he could create a symmetrical pattern. The line made by connecting these numbers is called a *magic line*. (Actually, in mathematical terms it is not a line, but a series of line segments and endpoints.) Bragdon found that these designs were very appealing to the eye, so he incorporated them into the decoration of the buildings he designed.

Look at the line created by the *lo shu* magic square from China. The *lo shu* magic square is considered one of the earliest examples of a magic square.

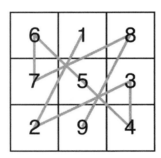 

To draw a magic line, first find the midpoint of each square in your magic square that contains a number. (If the midpoints are not obvious, you can find them in each square by drawing the diagonals between opposing corners.)

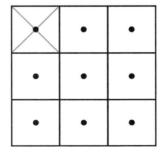

Then, working midpoint to midpoint, draw a line that begins in the square of the smallest number in the magic square and connect the line consecutively to each larger number.

What kind of design does the magic line make? Is it symmetrical? If so, how?

reate a magic line from each of the magic squares below. To make the most perfect magic line, use the point in each square as the beginning and ending of each consecutive line segment. The magic square on the left is from an ancient Tibetan seal. The magic square on the right is from an engraving called *Melancholia I* made by artist Albrecht Durer in 1514.

4·	9·	2·
3·	5·	7·
8·	1·	6·

1·6	3·	2·	1·3
5·	1·0	1·1	8·
9·	6·	7·	1·2
4·	1·5	1·4	1·

(1) What kind of symmetry do you see in these magic lines? How many degrees do you need to rotate the lines to find the symmetry?

Benjamin Franklin was also fascinated by magic squares. Here is a large one he created. Draw the magic line of this square.

5·2	6·1	4·	1·3	2·0	2·9	3·6	4·5
1·4	3·	6·2	5·1	4·6	3·5	3·0	1·9
5·3	6·0	5·	1·2	2·1	2·8	3·7	4·4
1·1	6·	5·9	5·4	4·3	3·8	2·7	2·2
5·5	5·8	7·	1·0	2·3	2·6	3·9	4·2
9·	8·	5·7	5·6	4·1	4·0	2·5	2·4
5·0	6·3	2·	1·5	1·8	3·1	3·4	4·7
1·6	1·	6·4	4·9	4·8	3·3	3·2	1·7

(2) What kind of symmetry do you see?

AND ANOTHER THING..............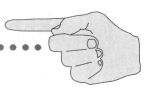

Look at one of the magic line designs you made. Color alternating spaces with black and another color of your choice. Can you see how Claude Bragdon might have found such a design useful in ornamenting his buildings?

8 NEEDLE AND THREAD GEOMETRY

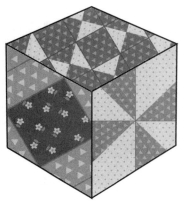

The art of quilting dates back centuries. Thrifty pioneer women turned scraps of rag and old neckties into useful coverlets and pillows that were often works of art in their own right. In the process of all this practical, yet creative, energy these women explored the dynamics of the relationship between geometric shapes to create visual effects that have held their appeal for generations. They pieced together congruent shapes or entirely different shapes in unique ways to give rise to formation after formation, design after design. Often the geometry of the quilt was a matter of pattern within pattern within pattern.

Probably the most basic shape to the creation of interesting quilt design is the triangle—and a right triangle, at that. This is because it is this triangle that is located within the basic quilt shape, the square. In addition to triangles, shapes such as squares and parallelograms such as diamonds, rectangles, and hexagons have found their way into the quilt square, or *block*, as it is sometimes called.

One very popular and durable design that shows us how shapes exist within shapes is the variable or eight-point star. This is the basic eight-point star design.

Creative quiltmakers have invented almost unlimited variations on this basic design. At left is an example of internal geometric shapes turning the simple eight-point star into so much more.

YOUR TURN

hile quilt designs are found primarily in fabric quilts, quilt designs make beautiful
recycled paper and pen-and-ink art. Use the square here to experiment with creation
of larger designs and shapes from the triangles found with this square. Once you have
created a geometric design you like (you may want to experiment on additional pieces of
graph paper), color the design or cut it out of used wrapping paper and glue the design to
another piece of paper to create a paper quilt block work of art. You may want to use
some of the design ideas on page 20 to inspire your thinking.

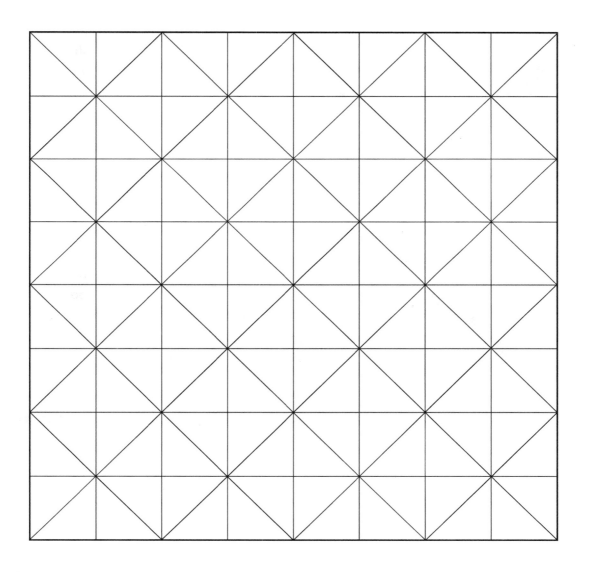

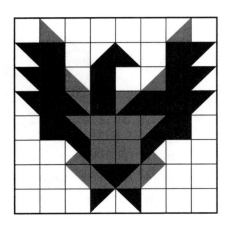

After you have experimented with geometric designs, you might want to explore using the quilt square and its internal shapes as the basis for an animal or human design such as that shown here.

Below are other traditional quilt block patterns that may interest you.

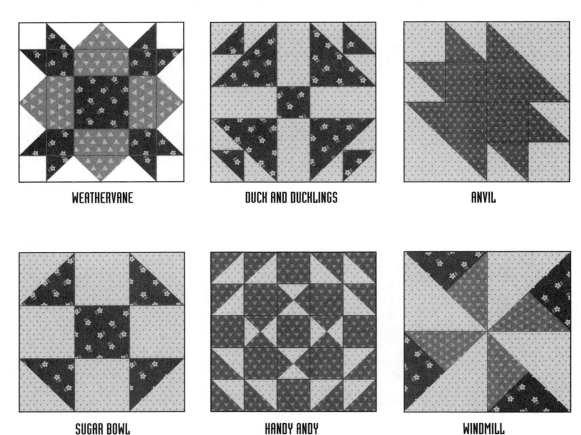

WEATHERVANE   DUCK AND DUCKLINGS   ANVIL

SUGAR BOWL   HANDY ANDY   WINDMILL

AND ANOTHER THING

The Japanese explore the geometric designs inherent in the sphere through the creation of *temari* balls. These small (2- to 2 1/2-inch-diameter) balls decorated with geometric thread designs begin as a small piece of bundled fabric. The artist winds thread tightly around the fabric until a spherical shape appears. The artist then divides the sphere into quadrants by lines of thread on which he or she creates a geometric design. Mothers and grandmothers made temari balls for children to use in games like kickball and handball.

TO MAKE A TEMARI BALL, FOLLOW THESE INSTRUCTIONS:

Materials:

▶ 6-by-6-inch square piece of fabric
▶ Ball of yarn
▶ Embroidery needle
▶ Embroidery thread (various colors)

1. Bundle the fabric and wrap yarn around it until you have made a ball. (Alternatively, you can begin with a hard foam ball that you wrap with yarn.) Once the ball is the size and shape you want, cut the yarn. Wrap the end of the yarn around another piece of yarn in the ball and tie it off.

2. Think of the ball as a globe with the north and south poles at the top and bottom of this sphere, respectively. With a needle and embroidery thread, sew the thread so that it cuts the sphere into two hemispheres going through the north and south points. Then, sew another thread dividing the sphere in half again at a 90-degree angle from the first thread. Now the sphere is divided into four equal segments or spaces north to south. Next, sew on an equator line that cuts the sphere in half going east-west.

3. Use these basic reference points to create a geometric design on the sphere. Use brightly colored thread to sew lines onto your sphere, crossing your basic lines until the entire ball is covered.

9 STRAIGHT TO THE ARC

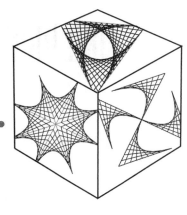

An arc is a curved line that is part of the circumference of a circle. Is there a way to turn straight lines into an arc? Can you turn straight lines into a parabola? A parabola is a special kind of curve.

Here's a bit of dot-to-dot fun. Each of the figures below shows you how you can make curved lines from connecting the dots on the rays of an angle. Here's what you do for each angle you explore:

1. Draw each of the angles shown in the figures below. Make one ray of each horizontal on your paper. For each angle, follow the directions below.

2. Place dots at regular intervals on each ray of the angle. The placement of the dots on each line segment or ray is the same.

3. Connect the dot farthest from the intersection on the top ray to the dot closest to the intersection on the horizontal ray.

4. Continue drawing line segments. On the top ray, as you move to a dot closer to the intersection to start a line segment, connect it to a dot on the opposing ray that is one point further from the intersection.

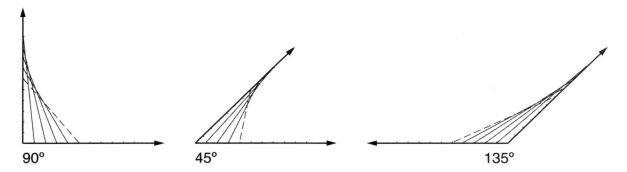

90°   45°   135°

(1) What do you notice about the shape you create from the lines? In which case do you create the arc of a circle?

Try something else. Draw your 90-degree angle again and this time place dots at irregular intervals on each ray. When you connect them in the same manner as you did before, what do you see?

YOUR TURN ························

You can create endless designs connecting dots with lines on geometric shapes. For example, you can make a circle. Use the four 90-degree angles in the corners of a square to make a circle as shown below at the left. Or, use the sides of an equilateral triangle to make an attractive design as shown below at right.

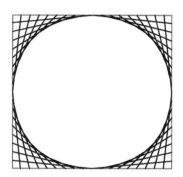
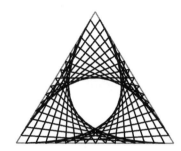

TO MAKE A DRAMATIC IMAGE OF ONE OF THESE DESIGNS, YOU NEED THE FOLLOWING:

► Black paper

► ¼-inch-square graph paper

► Glue or tape

► Ruler

► Pencil

► Needle

► Bright-colored embroidery thread

Draw a square, equilateral triangle, or other polygonal shape of your choice on a 6-by-6-inch sheet of graph paper and draw your design with ruler and pencil. Thread your needle and knot it. Glue the graph paper to the back of your black paper. Begin by pushing your knotted thread through the graph paper and onto the black paper to make your design. Now, needle your way to a dramatic geometric picture.

AND ANOTHER THING ·····················

If you make an arc or curve for each angle of a square or equilateral triangle, you can do a symmetry search of your design. Look at the figures on page 23.

(2) In the square, can you find reflective and/or rotational symmetry?

(3) In the triangle, what kind of symmetry can you find?

You might want to try to explore a more complicated design with your needle and thread. Look at the designs below.

(4) What kinds of symmetry do you find?

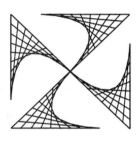

What kind of symmetry do you find in the designs you make?

10 ONE SIDE OR TWO—THE PAPER BAND MEETS AUGUST MÖBIUS

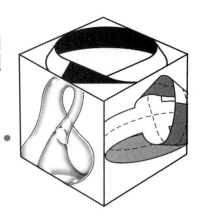

August Möbius was an astronomer who also studied mathematics, particularly geometry. He was very interested in a field of geometry called *topology*. Topologists are interested in finding out what makes stretchable and shrinkable objects alike and different from one another.

Möbius made some interesting discoveries when playing with a simple paper band. He explored whether an object has one side or two.

Take a strip of paper. How many sides does it have? Tape the strip into a ring. Trace around the band until you get to your starting point. Do you have a line on one side of the band only? Now trace the other side in the same way with a different-colored pencil.

(1)   How many sides does the band have?

PAPER STRIP   PAPER BAND

Take another strip of paper and put a half-twist into it before you tape it into a ring. Trace a line around the ring until you return to your starting point.

(2)   How many sides does this band have?

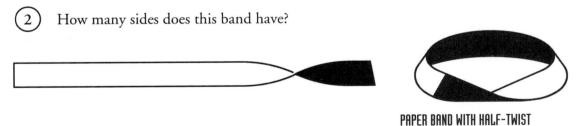

PAPER BAND WITH HALF-TWIST

Möbius's discovery has been valuable for users of machinery. Use of half-twisted rubber bands in machinery helps bands to wear more evenly as they go round and round, connecting machinery parts and doing their job. It gives the bands a longer useful life. This reduces the machine owner's cost of doing business.

Make three paper bands: a band (paper band with a half-twist), a band with no twist, and a band with a whole twist. Take one strip, the untwisted band, and the whole twisted band, and cut them in half down the middle, lengthwise, of each band. What do you find? What happens to each band? Trace a path with a pencil on each resulting band. What do you find?

Now, cut the band in half again. Trace the path of the band.

(3)   How many sides does it have now?

Make a band using this cross shape in the dimensions shown below. Tape one pair of opposing legs together with no twist and the other pair together with a half-twist.

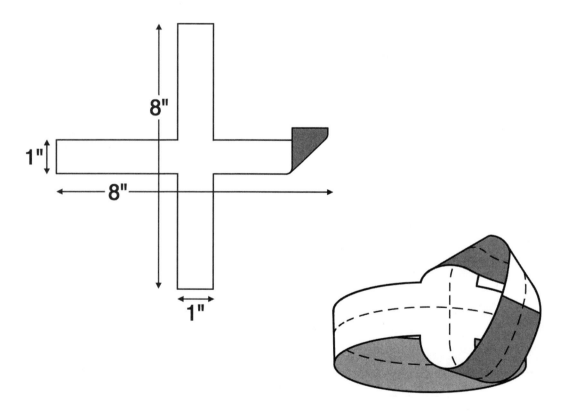

Trace a pencil path on this band. How many sides does it have? How is it like the strip? How is it different? Cut the band apart on the midline. What happens?

Make the cross shape again with paper. This time attach each pair of opposing legs with a half twist.

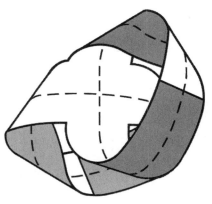

What happens when you trace the path? Cut this shape in half down the middle, or midline. What do you get? Try variations. What happens if you put a whole twist in the cross?

AND ANOTHER THING..................

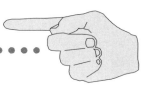

The Klein bottle is a three-dimensional version of the strip. It is a bottle with only one side. Here is how it looks.

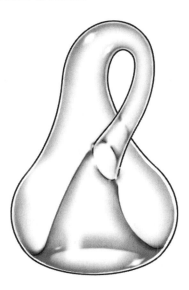

If you poured water into this bottle, what would happen? Where would the water go? Can you figure out how to make a Klein bottle from paper? If you cut a Klein bottle in half the right way, you get two strips.

(4) Can you figure out where to cut the bottle to turn it into two strips?

11 GENUS OF A SURFACE— THE TORUS

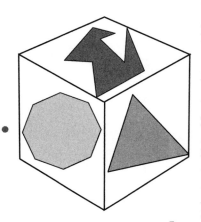

Is a

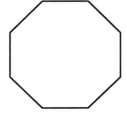

equivalent to a

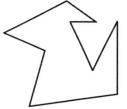

or?

In topology, the answer is yes. If that seems strange to you, it is because we tend to think of shapes on rigid, inflexible surfaces such as paper. In topology, mathematicians look at the features of an object as if the object can be stretched and pulled.

One way a topologist labels an object is by how many holes it contains. An object with no holes, such as a Frisbee or paper plate is a genus zero object. The genus of a surface or object is a number representing the maximum number of cuts you can make through the object without cutting it into more than one piece. Any cut through a genus zero object—that is, any cut that is not a hole, but cuts it into two pieces—gives you two new objects to evaluate. If an object has one hole in it, it is a genus one object because it has one hole. By making one cut in the object, you can turn it into a genus zero object.

A topologist looks at objects in this way because it helps him or her decide how objects that look different might really be the same, topologically speaking. For example, is a donut the same as a coffee mug? To a topologist, they are the same because both objects are genus one. They each have one hole and the same number of surfaces.

Let's look at the donut and coffee cup problem to see how the donut and coffee cup are alike.

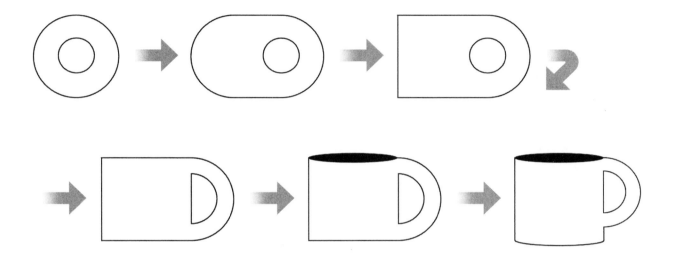

YOUR TURN  ●●●●●●●●●●●●●●●●●●●●●●●●●

To explore the topology of a donut, you will need the following:

▶ A piece of a balloon

▶ Clay

Let's begin with an examination of a piece of balloon. A topologist is not concerned with how big or what shape an object is. Cut a square from the piece of balloon. You can stretch it to make it bigger or twist it any way you want, but from a topologist's point of view, it still has the same properties.

Now, let's see how this works with a lump of clay. To a topologist, a donut and a coffee cup are equivalent because, while their shapes may be different, they have the same number of holes and the same number of surfaces. Explore this by doing the following:

1. Make a donut shape from your clay.

2. Gently change the shape of the donut by pushing the hole to one side of the donut. This hole will become your handle.

3. Take the large part of the clay and press it into a bowl or cup shape with the "hole handle" attached.

Try to do all these steps without tearing the clay into pieces. If you take off a hunk of clay to make the changes, you change the lump into two forms. You have made a "cut" and now you have two different objects to analyze topologically—you have to figure out the genus of the new objects you have made. Before you go to this next step, try to find the coffee cup in the donut without cutting or tearing the lump of clay.

AND ANOTHER THING  ●●●●●●●●●●●●●●●●●

Take a genus survey of objects around your home. Find out what genus number they have. Consider a bolt, nut, bowl, slotted spoon, book, and a plastic six-pack holder. Or, do a genus survey of the letters of the alphabet, grouping them into groups of genus zero, one, two, and so on. Graph your results. Remember, you are looking for holes.

12 IT'S OPTICALLY ILLUSIONING

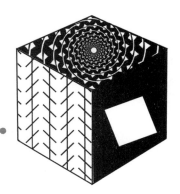

In mathematics, precision and accuracy are important. You cannot always trust your eyes when you look at geometric shapes, however, because they can play tricks on you. This whole idea of our eyes playing tricks on us became very interesting to the thinkers and researchers of the late 1800s. They wondered why we see things that aren't there.

These "tricks" are called *optical illusions*. In the 1800s, psychologists and physicists began to research, do studies, and write articles about the *what* and *why* of optical illusions. They found that optical illusions happen for several reasons. One has to do with the shape of our eyes. In other optical illusions, our minds have a certain response to the visual stimuli that makes us expect to see something that is not there. In some cases, the shape of our eyes works together with our minds to make the illusion.

Let's investigate some optical illusions that involve mathematical shapes. The following illusion is famous and is the illusion that may have touched off the interest of nineteenth-century thinkers into the subject. It was investigated by Johann Zollner, who lived from 1834 to 1882, and is sometimes called "Zollner's illusion." It consists of straight parallel lines and short angled lines.

Are the longer lines straight? Are they parallel? Measure the spaces between the long lines with a ruler to find out.

Another famous illusion is called the *twisted cord effect*. Is this picture a spiral or is it a series of circles? Trace the curved shapes with your finger to find out.

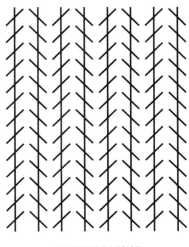

ZOLLNER'S ILLUSION

TWISTED CORD EFFECT

31

YOUR TURN

Make your own versions of Zollner's illusion. Try making the short angled lines at 60- and 120-degree positions. Draw the illusion again and make the angle of the short lines at 30- and 150-degree positions. Show your illusion samples to others. Does the change in angle of the short lines alter how people see the long lines?

Let's explore some other illusions involving shapes. Try the famous curved-shaped illusion. It is sometimes called the *convergence/divergence illusion* because our eyes are drawn in and then out, causing us to see the images in a certain way. Cut two versions of this shape, exactly the same size, from paper. Put one above the other. Does one look bigger? Which one? Place them side by side. Does one look bigger?

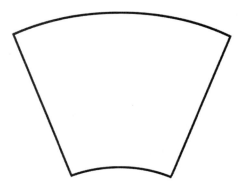

Now try another kind of size illusion. Draw a right triangle on graph paper. Make the vertical leg of the right angle 6 inches long and the horizontal leg 3 1/2 inches long. Draw two quarter-sized circles in the triangle—one at the top end of the 6-inch leg and one an inch or so above the 3 1/2 -inch leg. Does one circle look bigger? Try this one on your friends. Try this illusion again and move the circles into different places. How does this affect what you see?

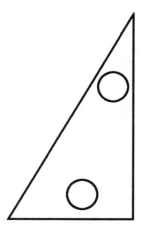

Make two 6-by-6-inch squares—one from black paper and one from white. Then make two 2-by-2-inch squares—one from black paper and one from white. Place the small squares on the square of the opposite color. Which square looks bigger?

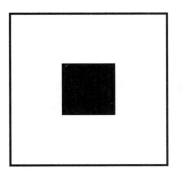 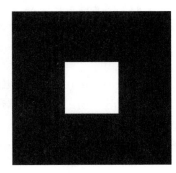

AND ANOTHER THING . . . .

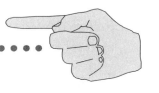

Find out how color affects optical illusions. Do some of the illusions above and add color to them. Try different-colored circles in the triangle. Try making the small 2-by-2-inch squares in red and blue and yellow and purple. How does color affect what you see? Because you know the sizes of the shapes you cut, try these illusions on your family and friends. Record the results. Can you think of some other illusions to try? Can you research some other illusions in books and change them a little, such as by adding color, to test them out on your friends?

13 A STRAW HOUSE

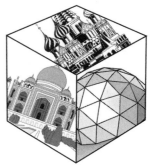

The history of architecture is the history of the use of mathematics to create larger, more dynamic, more useful, and more beautiful structures. To build the ancient pyramids of Egypt and Mexico, planners had to consider triangles, squares, polyhedra, and angles and use all their best estimation skills. In designing the great cathedrals of medieval Europe, architects analyzed arcs, angles, geometric shapes of many types, and proportional symmetry with mathematical formulae to determine how to use the center of gravity to create high, vaulted ceilings.

More recently, Frank Gehry and other architects have added mathematical questions to the discussion, such as how to use the parabola, spirals, and structural designs from nature in human-made public buildings. For example, inventor and architect Buckminster Fuller used spheres, angles, triangles, arcs, and other geometric concepts in his geodesic domes.

To begin to understand the integrated relationship between mathematics and architecture, take a survey of buildings in your own community. Look at homes, churches, synagogues, temples, public buildings, and local archeological sites. Look for framing structures, doorway shapes, windows, light openings, and stairways. Can you find polygons, polyhedra, arcs, spheres, spirals, and symmetry in these structures?

PARTHENON, ATHENS, GREECE

CHURCH OF NOTRE DAME DE LA GRANDE, POITIERS, FRANCE

TAJ MAHAL, AGRA, INDIA

ST. BASIL'S CATHEDRAL, MOSCOW, RUSSIA

YOUR TURN

To explore the effects of changing angles, proportion, and geometric shapes in architecture, try a building project of your own.

You will need these supplies:

- ▶ Plastic straws
- ▶ Scissors
- ▶ Straight pins, paper clips, and/or pipe cleaners

Using your knowledge of structures and their geometric foundations, build a tower. You can cut the straws to any length you choose. You can hook them together using straight pins, unbent paper clips (inserted into straws), and pipe cleaners (inserted single- or double-thickness into straws as needed to make them hold together).

What is the tallest structure you can build with a 15-centimeter diameter base? a 30-centimeter diameter base? What is the optimal length for your straws?

Now, try something a little different. The dome in the picture below is known as a *geodesic* dome. Can you build a geodesic dome with a 20-centimeter base? a 30-centimeter base? The dome structure has no interior support and can be made from triangular shapes.

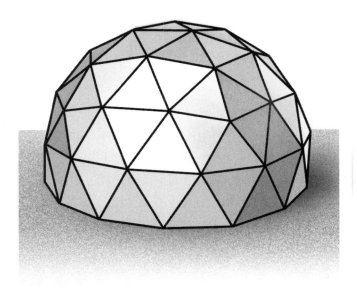

AND ANOTHER THING ....

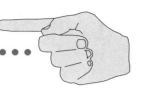

Different shapes and sizes of structures can hold different weight, or *load*. For example, columns might hold up the weight of a roof structure. Greek architects created three different styles of architectural design called Doric, Ionic, and Corinthian. Below is an example of the column top used in each design:

DORIC   IONIC   CORINTHIAN

Try your hand at designing paper columns. To make your columns, use 8 ½-by-11-inch paper. Roll it and test the strength of your pillar designs by piling on weight, or load— perhaps one paperback book at a time.

Change the diameter of your columns. Then change the length. How does this affect strength? Accordion-pleat the paper and make a column. Fold a square column. How do these changes affect pillar strength?

14 THE GEOMETRIC FOLD— SHAPES WITHIN SHAPES

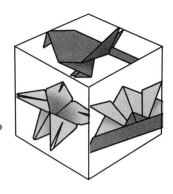

The simple act of folding paper can yield interesting mathematical results. One of the best known and most highly refined types of paper folding is the Japanese art of origami. Purists make their designs solely from folding the paper with no help from scissors or glue.

Historians believe that the art of origami dates back almost 1,500 years to the sixth century A.D., when Buddhist monks first brought paper to Japan from China. Paper was very expensive and valuable because it was still difficult to manufacture. As such, it was treated with care and great respect. Folded creations were made from it and became an important part of different Japanese ceremonies. Since those early beginnings, the art of origami has been shared and passed down from one generation to the next. Today, people around the world enjoy the craft of paper folding.

The basic paper shape is a square. Why a square? Fold a square piece of paper in half. Unfold it and fold it in half again the other way. By making these two folds, each of which divides the square into symmetrical halves, you create four congruent squares, or squares of the same size. Now fold the square diagonally so that you make a triangle when you fold the square in half. Unfold the paper and fold it in half the other way. These two folds divide the square into symmetrical triangles. What kind of angles do your half-folds make? The square allows the origami artist to make many folds that result in right angles. Do you think this has something to do with why the square is the basic shape from which other shapes are made?

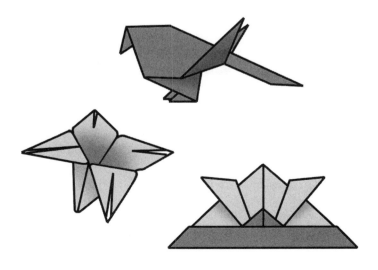

YOUR TURN ·

The art of origami offers many opportunities to explore different geometric shapes and their relationships. Fold a design of your choice or use these origami directions.

OLD UPRIGHT PIANO ·

1. Hold the paper square as shown below on the left. Fold the paper in half so that points 1 and 2 fall on points 3 and 4, respectively. Fold the paper in half again so that points 5 and 1 fall on points 6 and 4, respectively. Unfold.

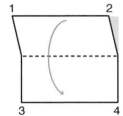 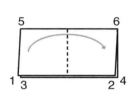 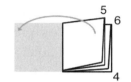

2. Fold points 5 and 1 to the center line and points 6 and 4 to the center line.

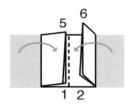

3. You are now holding a square shape with two rectangular flaps on the front. With the left flap, fold a triangle so that point 5 is on the outer fold of the left side of the square. Repeat with the right flap so that point 6 is on the outer fold of the right side of the square.

4. Open the flap at point 5 and flatten it to make a simple house shape. Do the same with the flap at point 6.

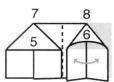

5. Fold the top edge that runs between points 7 and 8 down to the bottom of the two triangles.

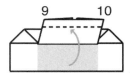 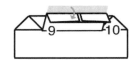

6. Fold up the top bottom flap (edge runs between points 9 and 10) so that the crease is at the bottom of the triangles. You have a rectangular flap. Now take the top of this flap (between points 9 and 10) and fold it toward you a quarter of the way down the width of the rectangular flap.

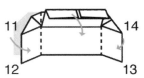

7. Fold in the two sides (running from points 11 to 12 and from points 13 to 14) to the center line and open to a 90-degree angle. See the picture below. Let the top rectangular flap down. Do you see the old upright piano?

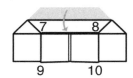

1. Hold the paper so that there is a point at the top. Fold point 1 to point 2. Now you have a triangle with the fold at the top. Fold the triangle in half and unfold it.

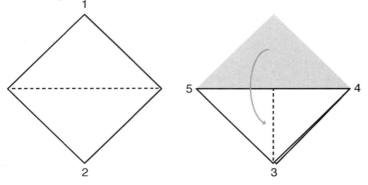

2. Hold the triangle at point 3 and open the paper at point 4. Press the fold line that ends in point 4 to the center fold of the triangle to meet point 3 and press flat. Repeat with point 5.

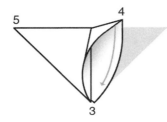

3. Now you have a diamond shape. Hold it with the open end down. Fold the top layer of paper at points 1 and 2 to the center line and press.

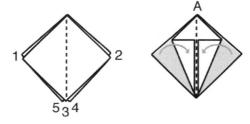

4. Fold point A down toward you. Unfold points 1, 2, and A.

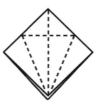

5. Lift the top layer of point B to make a long diamond. This is a bit confusing, but it works. Look at the picture carefully. Turn the object over and lift up the remaining layer of point B to make a second long diamond. The diamond has two "legs" at the bottom.

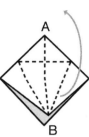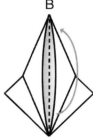

6. Hold the diamond so the two legs are facing down. Fold the two legs (points C and D) up as shown. Then unfold them.

7. Fold point C inside itself and squash it. Do the same with point D.

8. Fold the tip of point D down to make the bird's head. Look inside the bird between the two large triangles. There is a point. Fold the outer flaps at points E and F down on each side so that the crease line is at the inner point.

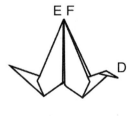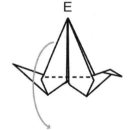

1. Hold the square so a point is at the top. Fold point 1 down to point 2. Then fold point 3 to point 4 and unfold.

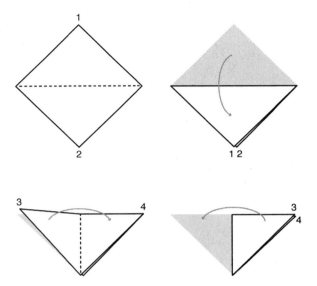

2. Fold point 3 to point 2 and press. Fold point 4 to point 2 and press. You have a diamond shape with the opening at the bottom.

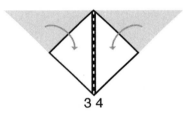

3. Turn the open point of the diamond to the top. Fold the left triangle at point 1 down so that the tip extends below the diamond shape. This is flap 4. Repeat with the other side. This is flap 3. These are the buzzer's wings.

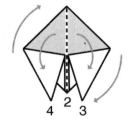

4. Take the top flap at point 1a and fold it down so that the crease line is about ³/₄ of the way from point 1 to the center fold of the diamond.

5. Turn the buzzer over. Fold point 1b down about three-quarters of the way to the center line (even with the fold line on the opposite side). Turn the buzzer back over.

6. Fold points 3 and 4 to the center line and turn back over. Turn corners at points 5 and 6 down at angles for eyes.

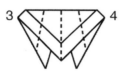

By folding a square of paper into a shape of your choice, you change the paper from a two-dimensional object into a three-dimensional one. Unfold your object and return it to the two-dimensional world.

It's time to search the creases on the paper to discover the mathematical secrets they hold. How many of the math principles below can you find illustrated on your origami square?

▶ Can you find lines of symmetry? How many can you find? Look from corner to corner and midpoint to midpoint. Why kind of symmetry do you see?

▶ What polygons do you find? How many triangles are there? Are they right-angle triangles, isosceles, and/or equilateral? How many quadrilaterals are there? Are there squares, rectangles, other parallelograms, rhombuses, trapeziums, or trapezoids? Can you find pentagons, hexagons and other multi-sided polygons? Can you find a polygon with eight sides? with twelve sides?

▶ *Congruent* shapes are those of the same shape and size. Can you find congruences? How many polygons of the exact same size and shape can you find?

▶ Can you find similarities? Are there small and large versions of the same triangle or quadrilateral? Measure the angles of the triangles to check this out. What is the ratio of the size of one similar figure to another?

▶ An iteration is a repetition of patterns within patterns. Can you find iterations?

In origami, you fold two-dimensional paper to make a three-dimensional object. Other paper-folding tricks are interesting in their own right.

1. Can you make a pentagon by folding a 1-by-10-inch paper strip?

   Can you fold a parabola? Take a sheet of 8 ½-by-11-inch paper. Place a dot about 2 inches from the edge on the 11-inch side. Now fold the sides of the paper again and again using the dot as the crease point but changing the angle of the paper, as shown in the picture below.

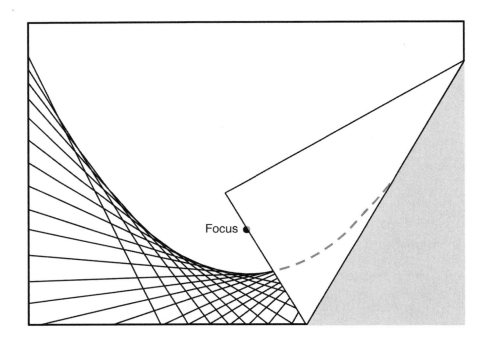

Focus

2. Can you fold an ellipse? Using what you learned about folding a parabola, can you figure out how to fold an ellipse using a circle-shaped sheet of paper?

15 PALINDROMES BY THE NUMBER

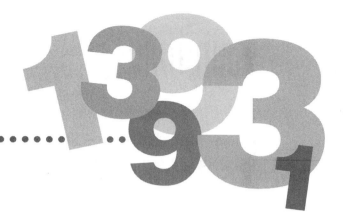

What is a palindrome? A palindrome is a phrase, word, or number that reads the same forward or backward. Here are examples of word palindromes:

MADAM   MOM   BOB   RACECAR

These are examples of sentence palindromes:

MADAM, I'M ADAM.   ABLE WAS I ERE I SAW ELBA.

A number palindrome is a number such as 12,321, which is the same no matter which direction you read it. When you play with numbers a bit, you discover that they have interesting properties. One of these properties gave rise to what mathematicians have called the *palindrome conjecture.* The conjecture said that you can make a number palindrome by adding any number to its inverse (for example, the inverse of 123 is 321), and then adding that sum to that sum's inverse and so on until the sum you get is a palindrome. It works like this:

$$
\begin{array}{r}
12,934 \\
+ \ 43,921 \\
\hline
56,855 \\
+ \ 55,865 \\
\hline
112,720 \\
+ \ 27,211 \\
\hline
139,931, \text{ which is a palindrome}
\end{array}
$$

This conjecture is true for a great many numbers. In fact, it takes less than twenty-four steps to reach a palindrome for most numbers. In the 1960s, a mathematician named Charles Trigg took a closer look at the old conjecture. He focused on numbers smaller than 10,000. He found that there were only 249 integers under 10,000 that did not yield a palindrome after 100 steps. He decided that they probably would not result in a palindrome no matter how many additions you did.

YOUR TURN

Choose some numbers of your own to test how many steps it takes to make a palindrome. Can you find any of the numbers that do not yield palindromes? You may not have that much spare time.

```
   12                  23,414
 + 21       + _____  + 41,432      + _____        + _____
```

AND ANOTHER THING

Here are some word and phrase palindromes: **bob, tot, dad, pop, pup, radar, pot top,** and **rotator.** Can you think of some others?

Can you make up some sentence palindromes? Here's a short one to get you going:

Tap Pat.

Don't worry if your palindrome sentences are a little strange. They are often a slice of nonsense.

16 NETWORKS NOT ON TELEVISION

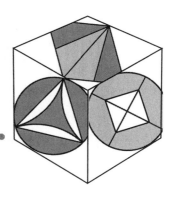

ou probably know what a maze is. It's something you can get lost in because of all the twists and turns and dead ends. But what's a network? Like a maze, a network is a path, but the path of a network is a series of lines and vertices (or places where the lines meet). You can also think of a network as a drawing or diagram of a problem. The problem is how to get from one place to another on the paths shown without going over part of the path more than once.

Some networks are *traversible*. This means that, by starting at one place on the network, you can trace the entire set of lines going through each arc (or line) only once without lifting your pencil. This is the problem of the network you must solve. While you can only go over each line once, you can cross the points or vertices as many times as you like.

Here are some networks you might know:

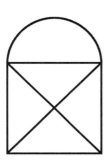
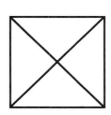
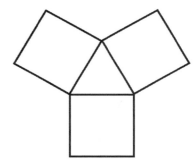

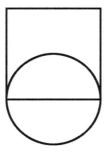
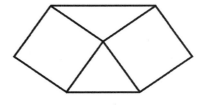

(1) Which of these networks is traversible?

YOUR TURN ·······················

Leonhard Euler was a famous Swiss mathematician who lived from 1707 to 1783. He studied the problem of networks to figure out why some were traversible and some were not. Can you figure out what Euler learned?

Look at the four networks below. Two are traversible. Two are not.

   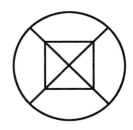

To figure out if a network might be traversible, Euler looked at the vertices of the network. A vertex is a point on the network at which lines and/or arcs intersect. Some vertices have an even number of arcs/lines passing through them. These are even vertices. Some vertices have an odd number of arcs/lines passing through them. These are odd vertices. How many odd and even vertices do each of the networks above have? How many odd vertices are in the traversible networks? How many are in the non-traversible networks? What did Euler find out about traversible networks?

He figured out that traversible networks can have no more than two odd vertices.

(2) Which of the networks above is traversible?

Draw your own network. Based upon the number of odd and even vertices your network has, can you predict whether or not it will be traversible?

AND ANOTHER THING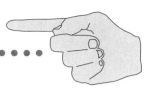

A famous network problem is the Konigsberg Bridge problem. In an old town named Konigsberg, which was located in Prussia (now part of Russia), the Pragel River forked around two islands. These islands were connected to the town by a series of seven bridges. Townspeople used to walk the path of the bridges and islands trying to figure out if there was a way to cross all the bridges just once in one continuous trip.

Is there a way to cross each bridge only once and traverse the entire path?

This problem came to the attention of Euler when he was working at the St. Petersburg Academy in St. Petersburg, Russia. While solving the problem of the Konigsberg Bridge path, he created a whole new field of math called *topology*.

③   Can you figure out what Euler's solution was? Is it possible to go over each bridge just once?

17 THE MATHEMATICS OF GAMES

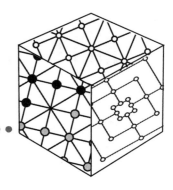

Many games involve a variety of mathematical issues and thinking skills that are useful in problem solving. This is especially true of the age-old games that have survived for centuries. Some of these games are merely games of chance in which probability is the center of all the game action. In other games, the more you play, the more you learn about the strategies involved in winning; it is this development of skill and the use of strategy that gives games a timeless appeal.

How is math really involved in the games we play—besides totaling up the score? Let's look at some of the ways.

PROBABILITY

Whether it is the roll of dice, the flip of a coin, or the spin of a dreidel, probability (or the chances of a certain result coming up) is a part, and sometimes all, of many games. You know that when you flip a coin, there is one chance out of two of a certain result. With the throw of a die (two die are dice), it is one out of six. With a dreidel spin, it is one out of four. (A dreidel is a four-sided top used in a game played during the Jewish holiday of Hanukkah.) In some games the probabilities become more complicated. This is true in the Native American game of sticks, in which there are six different objects thrown each time, with each object having two possible results.

LOGIC, STRATEGY, AND PROBLEM SOLVING

Playing games is not all, well, fun and games. Many of the games with the longest histories and broadest appeal involve lots of thinking and the use of strategy. Sometimes a player must think logically. Sometimes he or she must anticipate what an opponent will do and take that into account in his or her game plan. Sometimes games present a player with problems that he or she must solve in order to continue in the play.

NETWORKS

Many traditional games are played on a network-type design in which the playing pieces are moved from one vertex to another. A vertex is a point at which lines or angles intersect. Look at the board designs on the following page for two very old games.

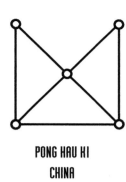

PONG HAU KI
CHINA

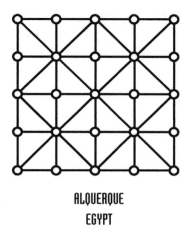

ALQUERQUE
EGYPT

Are these networks traversible or nontraversible? Do you think this aspect of the network makes any difference in the games played on these board designs? (See "Networks Not on Television," page 47, to learn about traversible networks.)

YOUR TURN

Take a look at some of the games you may have played. Choose one game and figure out the following:

▶ How does your game involve chance or the probability of one result versus another result?

▶ Do you use strategy to win the game, or is it simply a game of chance?

▶ If you use strategy to play, how does your strategy depend upon your opponent's skill or strategy in playing the game?

▶ Can you limit the effects of chance by developing a strong strategy?

Following are instructions for how to make and play games that have been played in some form since ancient times. Choose one of these games, play it and analyze it using the questions above.

PLAYING INSTRUCTIONS FOR GAMES OF THE WORLD

PONG HAU KI, A CHINESE VERSION OF TIC-TAC-TOE

Tic-tac-toe is a two-player game that most children learn to play. Pong Hau Ki is a Chinese version of this game. Like Tic-tac-toe, once you get the hang of it, it can be a very short game. The game is played with a board like the one shown here. The object is to use your markers to block any further move by your opponent.

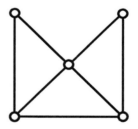

To make the game, make this design with a marker and a ruler on a stiff piece of paper or cardboard. Give each player two markers. Each player uses markers that are a different color from the other player's markers.

Playing Directions

1. Players take turns, and no one may pass a turn.

2. Play begins by first player taking a turn placing a marker on one of the vertices on the board. Then the second player places a marker.

3. If no one has won by the time all the markers are set, play continues with players taking turns moving markers on the lines to an empty spot or vertex.

4. The player who can block both his or her opponent's markers wins.

Try a variation on this game with the board modified as shown here. The rules are the same; two empty spots have simply been added on the vertical lines. How does the game work? Can one player win?

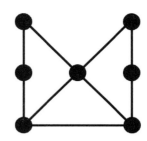

ALQUERQUE

Alquerque has its origins in the ancient Egyptian culture. Historians believe it traveled to northern Africa, and later the Moors brought it to Spain. The Spaniards took it to Mexico when they settled in the New World and the Native Americans of what would become the southwestern United States became Alquerque players.

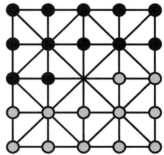

To play Alquerque, you will need to draw the board as shown at right on stiff paper. Each of the two players gets twelve markers. Each player's markers are a different color. To begin the game, players set up their pieces as shown here. The object of the game is to capture all of your opponent's pieces.

Playing Directions

1. Decide which player goes first. Players take turns moving their pieces to any adjacent empty spaces on the board.

2. A player can capture an opponent's piece(s) by jumping an adjacent piece and landing in an empty spot on the board. A series of jumps is permitted, just as in Checkers.

3. If a player fails to capture an opponent's piece(s) when he or she can, the opponent can capture or take the piece that should have jumped. Play continues until one player (the winner) has taken all of an opponent's pieces or one player has more pieces than the other and it becomes apparent that the other player has no more possible moves, or one player has put the other player in a position in which he or she cannot make a move. The game can end in a draw.

NINE MEN'S MORRIS

This board game is one of the oldest in the world. It is also known as Mill (meaning "row of three"). The first known version of this game was found on a temple roof in Egypt. Experts think it is more than 3,000 years old.

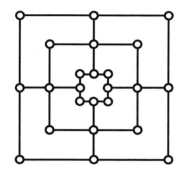

Here's how to make your own version of Nine Men's Morris. You will need to draw the design shown here on your cardboard with a marker. Use a ruler and place circles or dots on each corner—that is, where the lines intersect. Give nine markers of one color to each player, such as nine white beans to one and nine red beans to another.

Playing Directions

1. Each player gets nine playing pieces of a single color. Choose the player to go first.

2. Each player takes turns placing one piece on the board. Pieces may be placed where the lines intersect—that is, where the circles are. The object of the game is to place three pieces in a row on a line. This is called "making a mill." Once all pieces are placed, pieces are moved to adjacent open circles.

3. When a player makes a mill, he or she gets to take one of his opponent's pieces. No piece may be taken from an opponent's mill unless there are no other pieces to take.

4. The game ends when one player has only two pieces left. That player loses.

OVID'S GAME

This game is said to have been played by Ovid, the Roman poet. It may have been the forerunner of Mill or Nine Men's Morris. The game is for two players and uses a board like that shown here.

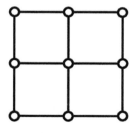

To make the game, you will need to draw the game board on stiff paper and locate six markers—three in one color and three in a second color. The object of the game is to line up three markers in a row.

Playing Directions

1. Players choose who will go first. Then players take alternating turns. No turns can be skipped.

2. First, the players take turns placing each of their pieces.

3. If no player has won by the time all the pieces are placed, players take turns moving their pieces to empty adjacent spots on the network until someone lines up three in a row. That player wins.

NIM

This game is thought to have originated in China. To make the game board, draw three horizontal lines on stiff paper. In the center, draw a vertical line as shown. Place three buttons, five buttons, and seven buttons on each horizontal line, respectively, to the left of the center line as shown below. The object of the game is to be the last player to be able to make a move.

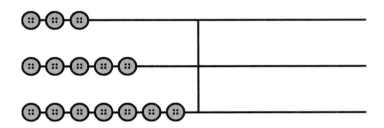

Playing Directions

1. Choose a player to go first. Players take turns. No player may skip a turn.

2. On each turn, a player chooses any number of buttons to move across the center line, but may only move buttons on one line each turn.

3. The winner is the player who makes the last move.

THE BEAN GAME

This game is similar to the ancient Chinese game called *nim*. To play this game, set up fifteen beans as shown below. The object of the game is to force your opponent to remove the last bean.

Playing Directions

1. Players take turns removing beans. Decide which player goes first.

2. On each turn, a player removes as many beans as he or she chooses, except that beans can be removed from only one row in each turn.

3. Can you figure out a strategy that virtually assures that you will win?

STICKS (OR WALNUTS OR PITS)

This game involving six playing pieces, each with two sides, has been played in Native American cultures from Mexico to Canada. It has been played with walnut halves, fruit pits, sticks, and small pottery disks.

To make the game equipment, you need six craft sticks or tongue depressors and twenty beans, buttons, pebbles or other counters. Decorate one side only of each stick and you are ready to play. While the game has been played with a variety of rules, here is a good set of rules with which to begin. The object of the game is to have the most counters at the end of play.

Playing Directions

1. Two players or two teams play. Each side begins with ten counters (or a number of your choice). The two sides take turns throwing the sticks.

2. Each throw is valued in this way:

   ▶ All sticks land faceup—player takes three counters from opponent
   ▶ All sticks land facedown—player takes two counters from opponent
   ▶ Sticks land three faceup and three facedown—player takes one counter from opponent
   ▶ Sticks land in any other combination—player takes no counters and loses no counters

   For a different version of the game, alternate throw values are:

   ▶ All sticks land faceup or facedown—player takes two counters from opponent
   ▶ Sticks land three faceup and three facedown—player takes one counter from opponent
   ▶ Sticks land in any other combination—player neither takes nor loses counters

KALAH

This is one of the oldest games in the world. It is thought to be at least 7,000 years old. It is very popular all over Africa, and one of its most well-known names in the United States is *mancala*.

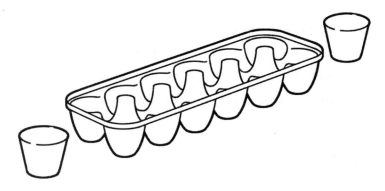

To make the game equipment, you will need one twelve-compartment egg carton (remove lid), two small cups, and thirty-six markers such as beans, small stones, or buttons. Set up the two cups at the ends of the egg carton. The cup to the right of each player is his kalah or special place. The object of the game is for a player to collect the most markers in his or her own kalah.

Playing Directions

1. Decide which player goes first. Then players take turns. Players may not pass a turn.

2. Place three markers in each of the board spaces. Place no markers in the kalahs.

3. To play, a player picks up all of the pieces in one of his or her six spaces (the six spaces on his or her side of the board) and then drops them, one by one, in each space moving counterclockwise. If the player arrives at the last space on his side of the board and still has markers, the player puts one in his or her kalah and continues counterclockwise placing them in his opponent's spaces. At no time does a player put markers in the opponent's kalah. If the player's last marker drops in a space on his or her side, that player gets another turn. If that space is empty, he or she gets to take all the markers out of his opponent's space directly opposite his empty space. These markers which he has captured go in his own kalah.

4. With these rules in mind, the players take turns. The game is over when all six spaces on either player's side of the board are empty. At this point, the player who has markers on his side gets to put those in his or her kalah. The player with the most markers in his or her kalah wins.

5. The game is made more difficult by starting play with more markers in each space. As you develop skill, try beginning play with four, five, or more markers in each space.

AND ANOTHER THING ⋯⋯⋯⋯⋯⋯⋯⋯

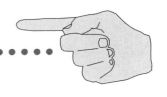

Make up a game of your own that interweaves probability, strategy, and networks. Create a playing board by developing a simple network. Decide how many players you need, what the playing pieces will be, and what a player must do to win. Can you involve probability in your game with dice or coin throws? Can heads and tails of a coin (signifying numbers like 1 and 2) tell players how to move or how many turns to take? Name your game and try it with your friends. What works well? What adjustments need to be made in the rules to make it work better?

18 PUZZLES AND GAMES— THE TOOTHPICK WAY

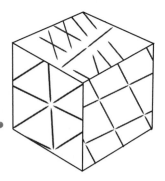

M any thinking skills go into solving math problems. The more advanced the mathematics, the more skills you need. You rely less on straight memorization and more on your ability to think clearly and logically. Many great mathematicians, scientists, and writers enjoyed puzzles and tricks. Lewis Carroll loved word games. Benjamin Franklin enjoyed making magic squares. Still others enjoyed puzzles such as toothpick and coin puzzles.

Successful puzzle-solving sometimes requires you to think in a logical way. Many puzzles distract the person puzzling them out with unnecessary information. To solve the puzzle, you must think in a straight line and avoid taking the wrong road because of assumptions you make. Sometimes puzzle-solving requires that you stop looking at the puzzle in the usual way and try to see it from a different perspective.

Toothpick puzzles allow you to exercise these skills and focus your thinking. Many of them are geometric in nature because the toothpick acts like a kind of line segment. Doing these puzzles exercises your skill in seeing the relationship between geometric designs and shapes. But, remember, not all toothpick puzzles involve geometric shapes.

Here are a couple of classics to get you started on good puzzle-solving thinking.

(1) *A Subtraction Puzzle*: Look at the fifteen toothpicks shown. Can you remove six to leave ten?

(2) *Lose That Square*: Look at the five squares formed by the toothpicks below. Can you move two toothpicks to turn five squares into four squares?

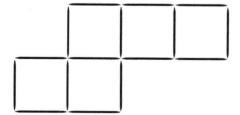

③ *A Triangular Puzzle*: Seven toothpicks make a triangle with a base of three toothpicks and two equal sides of two toothpicks each. Can you move three toothpicks to turn one triangle into three triangles? Hint: The three small triangles will be inside a quadrilateral with only two parallel sides.

④ *Triangles and Squares*: Can you make two squares and four triangles from eight toothpicks?

YOUR TURN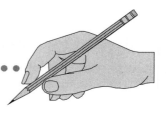

$\mathbf{H}$ave some fun solving these toothpick puzzles. You may need to think logically about how to move the toothpicks or you may need to try looking at the shapes in a new or different way in order to solve the puzzles.

(5) Can you move only one toothpick to make the following equations correct?

 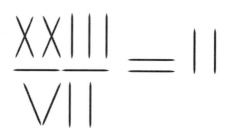

Some puzzles require you to think creatively about shapes and their relationships. Can you change one shape or group of shapes into another configuration of shapes?

(6) Find the relationship between geometric shapes. Arrange twelve toothpicks in a hexagon with six spokes. Move four toothpicks to create three triangles from the original design.

(7) Make a spiral from thirty-five toothpicks. Move four toothpicks of the spiral to make three squares.

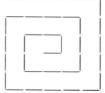

(8) Arrange twelve toothpicks in four connected squares. Move three toothpicks to create three squares.

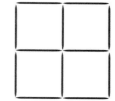

(9) Arrange twenty-four toothpicks in nine squares arranged in a 3-by-3 block. With twelve more toothpicks, create four more squares. Then, remove four toothpicks from your design to leave nine squares.

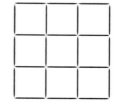

From *Math Amazements*, Copyright © Good Year Books. This page may be reproduced for classroom use only by the actual purchaser of the book. www.goodyearbooks.com

AND ANOTHER THING ....

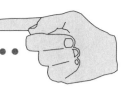

Now let's try a little coin fun.

(10) Can you turn this triangle upside-down by moving only three coins?

(11) Arrange six coins in a cross shape. Move one coin to form two rows, each of which has four coins.

(12) Arrange twelve coins in a square. Rearrange them to form another square with five coins on each side.

19 UNPUZZLING THE TANGRAM

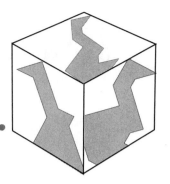

The tangram is a Chinese puzzle that became very popular in the 1800s. Tangram puzzles made from plastic and wood are available as toys today for people of all ages. One source of the tangram puzzle was a Chinese book from 1813. Lewis Carroll, the author of *Through the Looking-Glass*, and Edgar Allan Poe, the author of "The Tell-tale Heart," were both great tangram enthusiasts.

The tangram is a square that is divided into seven pieces. Five of the pieces are both isosceles and right triangles. An isosceles triangle is a triangle with two equal sides. If such a triangle has one 90-degree angle, it is also a right triangle. There are two large triangles (of the same size), two small triangles (of the same size), and one medium-sized triangle. In addition, there is one square and one parallelogram. Over the years, more than 1,600 shapes have been developed using the tangram puzzle to create images of animals, people, and objects.

The puzzle is made in the following manner:

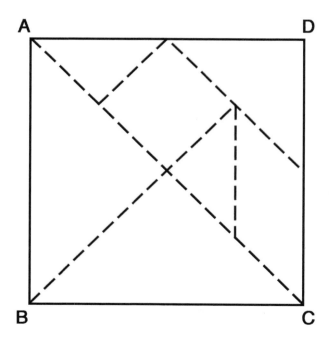

Obviously, you can fit the seven pieces into the shape of a square because that is the shape from which they originate. These same pieces can also be used to form the following polygonal shapes:

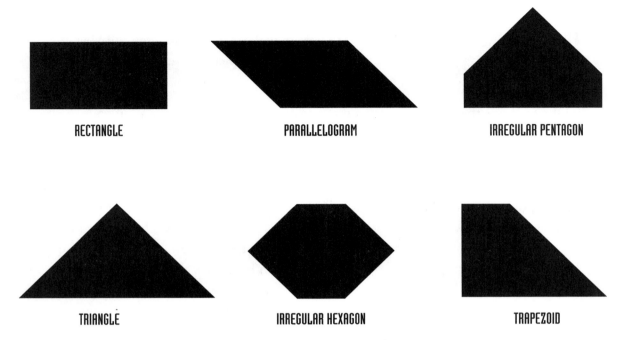

RECTANGLE PARALLELOGRAM IRREGULAR PENTAGON

TRIANGLE IRREGULAR HEXAGON TRAPEZOID

YOUR TURN ....

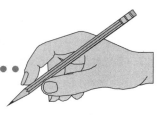

Begin with two 6-by-6-inch squares of paper to make your own tangram puzzle—one scratch paper on which to design the puzzle and one piece of sturdy paper or cardstock from which to make and cut your final puzzle.

(1) Examine the tangram puzzle on page 62. How can you recreate it on your square so that the pieces are the correct size?

First, think about what you know. You know how to make the two large isosceles right triangles located within the larger triangle ΔABC. Next, look at the puzzle and check out the relationships between the sides of the pieces that intersect the diagonal of the square. Measure them. How do the lengths of the sides relate to each other? Then, measure and find out about the relationships between the pieces that intersect $\overline{AD}$ and $\overline{DC}$ of the square. Does the intersection point cut each line segment in half? How does this information help you figure out how to draw the rest of the tangram puzzle? Now, using a ruler and pencil, measure and draw your own tangram (on scratch paper or directly on your sturdy paper, as you choose) and cut out the pieces.

(2) Can you put the pieces back together to create any of the polygons shown on the prior page?

(3) Can you make the computer desk shown here?

(4) Here are a few creations that reside in tangram zoological parks. Can you figure out how to make them?

NAPPING LIZARD

LONELY BEAR

WATCHFUL VULTURE

ICE-SKATING GOOSE

(5) Here are three more puzzles for you to try:

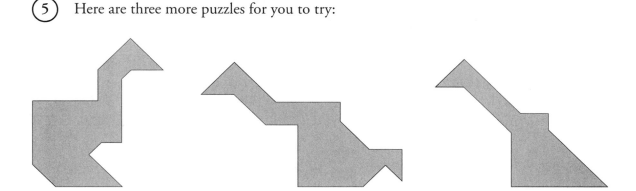

GOOSE   DINOSAUR   SEA MONSTER

Now try making up your own designs to stump your friends.

AND ANOTHER THING · · · · · · · · · · · · · · · · · ·

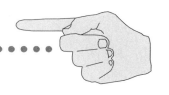

Making puzzles from geometric shapes has fascinated mathematicians throughout history. We know a little about the Chinese tangram, but Euclid, the famous Greek mathematician, was fascinated by the idea of dividing a rectangle into pieces that could make other geometric shapes. He was thinking and writing about this more than 2,000 years ago.

(6) Use what you know about basic geometric shapes to make up your own Euclid's rectangle puzzle. Consider using three isosceles right triangles, a square (divided in half), two parallelograms, and a *trapezium*. A trapezium is a quadrilateral with no parallel sides. Divide your puzzle into eight different shapes. What shapes or designs can you make from the pieces of your rectangle puzzle? Can you stump your friends? Can they stump you?

20 HOW THE EGYPTIANS MULTIPLIED

The ancient Egyptians were masters of architectural achievement and invention. They built the great pyramids. They developed the square sail to harness wind power for boating. They were able to develop a highly advanced and complex civilization despite use of very simple arithmetic. In fact, they really used only addition and subtraction. To multiply, they relied on addition.

Let's see how this worked. If an Egyptian wanted to find out how much 15 times 31 was—that is, the total of adding 31 fifteen times—he used a doubling technique. In our system of multiplication, we call the larger number the *multiplicand* and the smaller number the *multiplier*. In the problem, 31 x 15, 31 is the multiplicand and 15 is the multiplier.

This is how the Egyptians used the multiplicand and multiplier to do multiplication:

To solve the problem of 31 x 15, the Egyptians made two columns that they doubled like this:

Number of Multiplicands	Total
1	31
2	62
4	124
8	248
16	496

Starting with one multiplicand, they doubled the numbers in each column until some combination of numbers in the first column totaled the multiplier. In the first column above, 1 + 2 + 4 + 8 equals 15, or the multiplier. The Egyptians then added the numbers in the second column that were across from these numbers to find the solution to 31 x 15.

$$
\begin{array}{c}
31 \\
62 \\
124 \\
248 \\
\hline
465
\end{array}
\quad \text{or} \quad
\begin{array}{c}
31 \\
\times\ 15 \\
\hline
465
\end{array}
$$

YOUR TURN ·

Solve 43 x 9 using the Egyptian method.

1	43	
2	86	43
4	172	+ 344
8	344	387

Lines 1 and 8 total 9, the multiplier. Add the numbers in the second column that relate to 1 and 8 to find the answer to 43 x 9. This addition is done for you in the third column above with the answer in Egyptian multiplication being 387. Now you multiply 43 x 9 in your usual way to check the answer. Do you get the same answer?

Why does this work? It works because the Egyptian method is just another way of putting together or totaling the numbers you multiply. Look at the chart below.

Your Method	The Egyptian Method
43	43 > one 43
x 9	
387	+ 43
	43
	43
	43 ⟩ eight 43s
	43
	43
	43
	43
	387

Here are three more problems for you to try. Of course, you could multiply them the ordinary way. Do them instead using the Egyptian method.

17 x 12   23 x 16   16 x 13

Now try Egyptian multiplication with three problems of your own.

AND ANOTHER THING

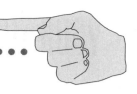

The Egyptians were a very practical people, so they only developed mathematical processes that they needed for the jobs that faced them. Perhaps this is why they only developed a very basic form of fractions. While they knew that fractions were parts of a whole, they never learned how to use any number but 1 as a numerator. For example, they understood $\frac{1}{2}$, but did not understand $\frac{4}{8}$. If they divided a whole into $\frac{7}{12}$, they could only write it as a sum of fractions with numerators of 1. Using this method, $\frac{7}{12}$ becomes $\frac{1}{2} + \frac{1}{12}$.

(1) What is $\frac{13}{16}$?

21 MAGIC SQUARE IN THE MAKING

A magic square is a group of numbers arranged in a certain way so that the numbers have an interesting property—that is, the sum of any row, column, or diagonal is the same. Historians believe the first magic square called *lo-shu* dates from ancient China before 2000 B.C. There is a legend about its discovery that says an emperor saw the special number sequence on the shell of a tortoise near the bank of the Yellow River.

The *lo-shu* magic square is made up of nine smaller squares, three across and three down. Each square includes one of the first nine whole numbers arranged in such a way that no matter which direction you add them, including diagonally, the numbers total 15.

Once you have a magic square, you can turn it into a new magic square by switching the position of complementary numbers. Complementary numbers are any pair of numbers that, when added together, equal the sum of the largest and smallest numbers in the square. For example, the sum of the largest and smallest number of the *lo-shu* square is 10 (1 + 9). Complementary numbers for this square are 2 and 8, 4 and 6, and 7 and 3. What happens if you only change some of the complementary numbers? Do you change the properties of the square if you do not change all the complementary numbers?

It takes a lot of time to make a magic square if you have to choose numbers by trial and error. This has caused many math enthusiasts through history who enjoyed magic squares to think about methods for making them. Benjamin Franklin was one such famous magic square fan. Another man, Antoine de LaLouvere (1600–1664), developed a method for making magic squares that is called the *stairstep method*. It works like this:

▶ Using the number sequence 1 through 9, begin with 1 in the top middle square.

▶ Place each subsequent number (going in consecutive order) in the space one line above and one space to the right. If this space is off the square, you find the corresponding place on your square and put it there.

▶ If that space is already filled, you put the next number directly below the number you just placed.

(The space for 7 is already filled so 7 goes into the block under 6, just as 4 went into the block under 3.)

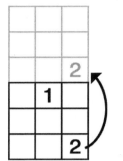
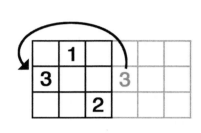
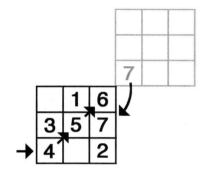

YOUR TURN

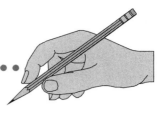

ry your hand at making your own magic square.

(1) Start with a five-by-five square using the numbers 1 through 25. If you are a kindred spirit with Benjamin Franklin, you might try to devise your own method. Otherwise, try the stairstep method on this five-by-five square.

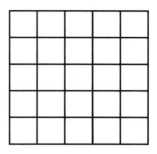

Make a new magic square from the one you just made by switching all the complementary numbers. Is this like simply flipping over the square?

(2) Now, is it possible to use the stairstep method to create a four-by-four magic square? Why? Why not?

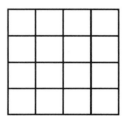

Can you figure out your own method for making a magic square with the numbers 1 through 16?

AND ANOTHER THING

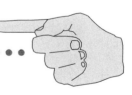

f you are tired of magic squares, how about trying your hand at an *antimagic square*? This is a number square in which each of the rows (horizontal, vertical, and diagonal) totals to a different number.

(3) Use the whole numbers 1 through 9 to make a three-by-three antimagic square.

22 THE FOUR-COLOR MAP PROBLEM

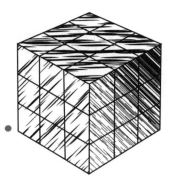

For centuries, mapmakers followed an unproven rule for maps drawn on flat surfaces or on spheres. The rule says that you need only four colors to differentiate adjacent countries or territories so that no two colors share a boundary; they can share vertices or corners only, a vertex being the point at which lines intersect. This was called the *four-color map problem* because, while this was every mapmaker's rule, no one had been able to prove that there wasn't a map somewhere that might require five colors.

Take a look at how the states are colored on a map of the United States. How many colors are used? Or, look at a globe and see how the countries on a continent are colored. How many colors are used?

While this four-color rule might not seem so difficult to prove, mathematicians have struggled for generations to prove it. In 1840, August Möbius raised the question of whether topologists could find a proof. In 1879, Arthur Kempe, an amateur mathematician, published a proof, but other mathematicians found an error in it. Kempe's proof did lead, though, to a proof that all maps could be colored with a minimum of five colors. It wasn't until 1976 and several thousand hours of computer time that a proof for the four-color map problem was found. There is still no proof that can be done simply with pencil and paper and there are still some mathematicians who dispute the validity of the 1976 proof.

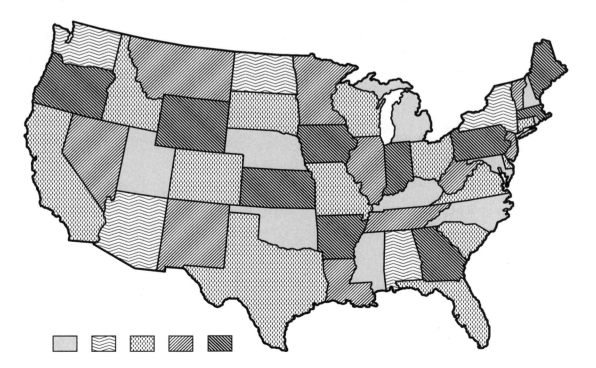

YOUR TURN ....

Try your hand at mapmaking to see how many different colors you need to color different regions so that no two regions with a common boundary have the same color. Always try to use the minimum number of colors possible. Try to use only four colors.

This is a map of part of the United States. What is the minimum number of colors you need to color it so that no two shared boundaries are the same color?

This is another map of part of the United States, but it has been divided as though there were even more states. What is the minimum number of colors necessary to color this map now?

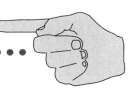
With the five-color theorem, mathematicians were able to prove that the most colors ever needed to properly color a map is five. Insert imaginary county lines into this map of Nebraska. Try to divide the state into territories so that you must use five colors. Is it possible? If you think your lines require five colors, test your lines on a few friends to see if they can color your map with four colors only.

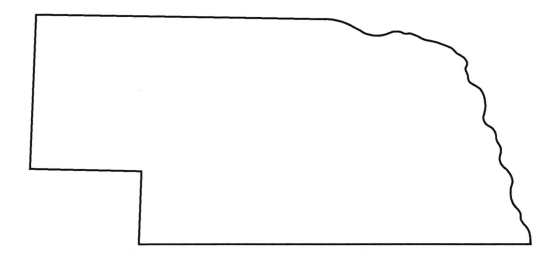

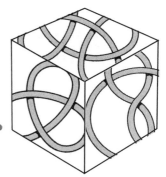

23 WHAT'S KNOT TO LIKE?

Most of us have been tying knots since we learned how to tie our shoes. How can it be, then, that knots are a very big topic of discussion and research among mathematicians who study the field of topology? Topology is the branch of mathematics in which we look at the properties or characteristics of objects that remain the same even when their shape is changed—that is, when they are stretched or shrunken.

Knot theory and resulting research into what makes a knot a knot is a new area of study in the field of topology—which itself is new to mathematics—but knots are very old. Early humans learned how to tie knots of different types to help do the work that needed doing. Some early Egyptians were rope stretchers, people who tied knots in ropes to help them find right angles so they could build pyramids and temples. Sailors have been using a variety of knots for centuries as they go about the business of sailing, docking, and anchoring their boats.

So what's not to like about knots? We all need them and we use them every day for many things, but what are they? To a mathematician, a knot is a piece of string or cord with loops and crosses that has no loose ends and that cannot, with pulling and tugging, be turned into a circle. So you might say that one thing a knot is *not* is a circle or "unknotted curve." Some mathematicians call a circle the "unknot" or the "zero knot," because it has no crossings that cannot be twisted or stretched away.

You can make a knot by looping and crossing a piece of string and then joining the ends of the string together. If this "closed curve" of string remains knotted no matter how you twist and pull it (so long as the string is not broken), then you have a knot. If you twist and pull on it and it dissolves into a circle, you have no knot.

For topologists, the simplest knot has three crossings and is called a *trefoil knot*. A left trefoil is shown on the left below and a right trefoil is shown on the right.

LEFT TREFOIL KNOT

RIGHT TREFOIL KNOT

Do you think they are the same? If you turn them over or manipulate them, can you make a right trefoil into a left or vice versa?

Early knot researchers had to make their own knots. They tied knots and analyzed them and by this process figured out which knots were like each other and which were not. Now mathematicians can use computers to analyze their knotty problems.

When you look at knots to see how they are alike and different from one another, you can look at a number of characteristics. How many crossings does each knot have? These are the places at which the string or cord crosses over itself. How many loops are there? Are the crossings under or over another part of the cord or string?

(1) Look at the left and right trefoil knots on page 75. Make samples of them with cord and tape the ends together. Are they the same knot? Can you manipulate one to turn it into the other?

(2) Look at the two knots below. Make models of them from stiff string or cord. Are they the same knot? Analyze the knots in terms of loops and crossings. How many crossings does each have?

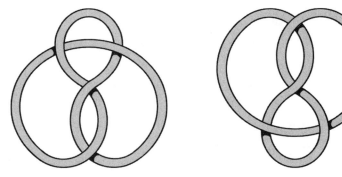

(3) Here are two more knots. Make them out of stiff string or cord and look at them closely. Are they the same knot? See how many crossings they have and what is alike and different about them.

Analyze the knots you have made. Are all of them true mathematical knots, or can you manipulate them so that they become circles?

AND ANOTHER THING

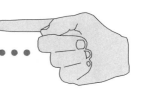

Let's try some other knotty activities. Here are some further knot explorations:

▶ Can you make a knot in a piece of string by unfolding your arms? Cross your arms and fold them on your chest. Now pick up the two ends of a piece of string, one end with each hand, and pull. What happens?

▶ Can you make a trefoil knot out of paper? Take a strip of paper and give it three half-twists and tape it together. Draw a line down one side and find out how many sides this object has. If your line meets the line where you began, the object has one side only. Now cut the loop in half right down the middle of the strip so that the cutting line is right in the middle of the edges the whole way around. Have you made a trefoil knot? Compare its crossings with your trefoil models.

24 JORDAN CURVE? CAN YOU THROW ONE?

Is a Jordan curve some kind of a pitch? And who is Jordan anyway? You may not think so now, but you do know what a Jordan curve is. It is a "closed curve." A closed curve is one that does not intersect itself. The simplest Jordan curve is a circle. The important feature of the closed curve is that it divides a plane into two distinct places—inside the curve and outside the curve. The curve was named after a French mathematician named Camille Jordan who lived from 1838 to 1921. He had the idea that any closed curve divides a plane into what's inside and what's outside of the curve. You might wonder what made Jordan so special that he got the circle named after him, because if you'd been around you could have easily told him this.

Well, Jordan wasn't looking only at circles. Both of the designs below are Jordan curves. In the circle at the left, point A is outside the curve and point B is inside the curve. In the circle it is easy to tell which is which. In the case of the very curvy closed curve at the right, it is much harder to tell.

(1) Can you figure out whether points A and B are inside or outside of the curve?

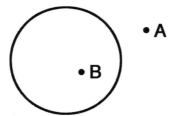

Jordan came up with some very complicated thinking about the ins and outs of the Jordan curve, and that's why he got to put his name on it. Let's find out more ourselves. For example, is there any way you can tell by looking at the closed curve on the right whether or not it is a Jordan curve?

YOUR TURN ....

Think about a circle on a piece of rubber that you could easily bend, twist, and stretch. No matter how the line of the circle changes through twisting the rubber, the circle is still there. If you had put point A inside the circle, it would still be inside the circle no matter how you twisted, turned, and stretched that rubber.

Not all Jordan curves are as simple as a circle or an ellipse. The curve below looks like a maze, but it is a Jordan curve. The characteristic that makes it a Jordan curve is that it divides the plane it is on into two distinct areas (inside and outside the curve).

2) Can you tell which dot is inside the curve? Is dot A? Is dot B?

To find out if a dot is inside or outside the curve without tracing your finger through the maze, pick a direction from the dot to a point well outside the curve. Do not move parallel to any side of the curve. Draw a line from the dot to the outside point. A dot is inside the curve if you can count an odd number of lines from the dot to the outside. A dot is outside the curve if you can count an even number of lines. Test this first on the Jordan curve illustrated above and then by drawing your own complicated Jordan curves. Does the rule work? Turn your Jordan curves into mazes by erasing a spot in the outside wall of your snake-like curve.

AND ANOTHER THING ....

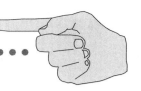

Can a Jordan curve be a *polygon*? A polygon is a closed figure made by three or more line segments. Because a line segment is different from a curve, then perhaps a polygon cannot in a strict sense be a curve, but mathematicians have talked about Jordan polygons. Look at the spiky, maze-like creation illustrated below. It is a many-sided polygon that looks like a maze, just like a Jordan curve can look like a maze.

(3) Will the method you used to find if dots are inside or outside of the figure work on this polygon? Try it, and see.

Did the method work for this polygon? Make up your own many-sided polygon and test the result. Does it always work? Now, turn your spiky creation into a maze.

25 AN A-MAZE-ING THING

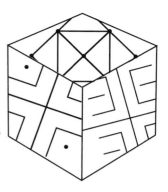

People of all ages do pencil-and-paper mazes. A maze is like a network in that it is a path, but in a maze the correct path leads you to your goal—either a place in the middle of the maze or a way out. To do the maze properly, you need to reach your goal without crossing any lines and without picking up your pencil.

Mazes have been around since the time of ancient cultures. One of the oldest known mazes is found on a coin from Knossos, Crete. Stone carvings in Ireland show mazes that are more than 3,000 years old. The Greeks loved mazes, and, according to one Greek myth, the Minotaur was housed in a maze. Mazes are found in African and Native American textile designs and in classic English gardens.

COIN DESIGN FROM ANCIENT CULTURE
AT KNOSSOS, CRETE

NATIVE AMERICAN NAVAJO
TEXTILE DESIGN

ENGLAND'S HAMPTON COURT GARDEN MAZE DESIGN

Can you find your way through these mazes?

There are many ways to design a maze. One way is to create a traversible network and turn it into a maze. A traversible network is one that, by starting at one place on the network, you can trace the entire set of lines going through each arc (or line) only once without lifting your pencil. (For more information about networks, see "Networks Not on Television," page 47.)

(1) Can you figure out a way to change these networks into mazes? Remember to include a goal and create some dead ends.

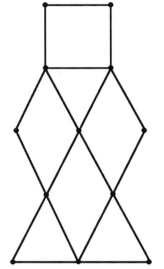

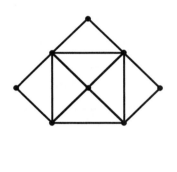

As you think about how to make your own maze, it sometimes helps to look at famous mazes. Look at the Hampton Court maze on page 81. What aspects of the maze are symmetrical? What aspects are asymmetrical?

(2) Can you build mazes from these basic symmetrical designs?

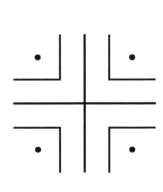

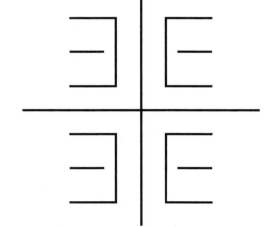

Build your maze by connecting the lines and dots to one another in some way.

AND ANOTHER THING ....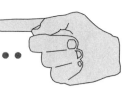

Look at the Knossos coin maze design on page 81.

(3) Can you figure out if the maze is a Jordan curve? (For information on Jordan Curves, see "Jordan Curve? Can You Throw One?" on page 78.)

26 IS IT PROBABLE?

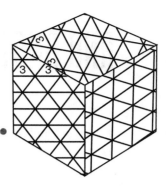

Western mathematicians first began to formally investigate probability in the 1600s. Probability is the study of chance. They began their studies with coins and dice because there were a limited number of results possible from the throw of coins and their experiments and observations would be easy to undertake. Think about it.

1. If you flip a coin, how many possible results are there? When you toss a die, how many possible results are there?

Blaise Pascal, who lived from 1623 to 1662, was a French mathematician and philosopher. Pascal created a number triangle that can be used to identify the probability of a certain outcome or result occurring when, for example, coins are tossed. It appears that, as early as the eleventh century, both a Chinese mathematician named Chu Shih Chieh and a Persian poet named Omar Khayyam were aware of and used the same triangle.

The triangle works as shown in this drawing:

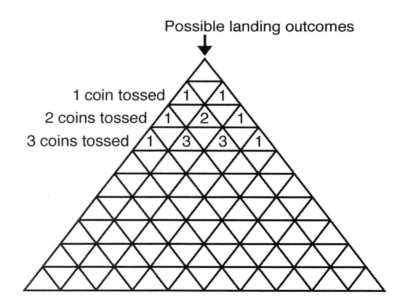

Possible landing outcomes

1 coin tossed 1 1
2 coins tossed 1 2 1
3 coins tossed 1 3 3 1

When you toss one coin, there is one chance the coin will fall heads and one chance the coin will fall tails. This means there is one out of two chances that, if you guess heads, the coin will fall heads. When you toss the coin twice, there is a one in four chance that you will get heads each time (that is, one chance that you will get two heads, one chance that you will get two tails, and two chances that you will get one head and one tail each time).

There are many patterns to the numbers that appear on Pascal's triangle. Look at the triangle below.

(2) Can you figure out the number of possible outcomes for each group of coins tossed?

If you can figure out the pattern of the numbers inside the triangle, the total outcomes are easy. As you look at the triangle below, remember that on the left side of the triangle the numbers tell you how many coins are tossed. On the right side of the triangle, the numbers tell you the total possible outcomes when you toss that number of coins. The numbers inside the big triangle tell you how coins will fall—for example, if you toss two coins, there is one chance both will fall heads, one chance that both will fall tails, and two chances that the coins will fall one head and one tail.

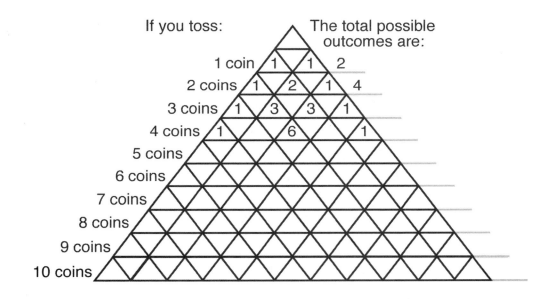

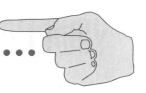

AND ANOTHER THING ·····················

Over and over again in mathematics, you can see how one number pattern relates to another. One such pattern, called the *Fibonacci number sequence,* is one in which a series of numbers is made by adding the two preceding numbers to get the next number. The series looks like this:

1, 1 (0 + 1), 2 (1 + 1), 3 (1 + 2), 5 (2 + 3), and so on

③ If you add the numbers on the diagonal lines drawn through Pascal's triangle, what number sequence do you find? Do this on the Pascal triangle you completed on page 85.

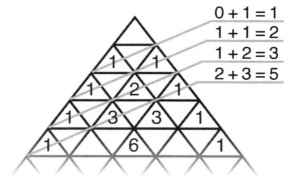

④ Can you find any other number patterns in the triangle?

27 CAN YOU GAUSS THE ANSWER?

Carl Friedrich Gauss was a famous mathematician who lived from 1777 to 1855. He discovered or found equations for many interesting mathematical principles. Finding equations that prove something about how numbers work is a big part of what mathematicians do. Their equations are not simply numbers that you add, divide, or multiply. Their equations are about the relationship between unknown numbers or about numbers that are different, depending upon the size or shape of a thing.

Gauss had amazing mathematical skills that were obvious even when he was very young. There is a story about his experiences at age 8 in elementary school that illustrates his prowess. One day his teacher asked the class to find the sum of all the numbers 1 through 100. While the other children sat there busily scratching out sums and adding more sums, Gauss looked off in a world of his own. His teacher scolded him to get to work to which he replied that he already had the answer. He had figured it out in his head.

(1) Can you figure out how Gauss did this problem in his head? If you figure out his method, you may be able to do it without pencil and paper, too.

Gauss explored many areas of mathematics. Although the normal distribution curve was discovered in the late 1600s, it was Gauss who developed one of those famous mathematical equations for it. The curve is now called the *Gaussian curve*. It looks like this and is the same curve that shows the possible outcomes when a pair of dice is thrown.

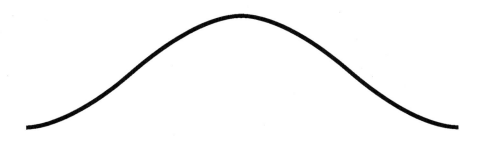

YOUR TURN 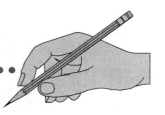 •••••••••••••••••••••

It's time to try your hand at some of those puzzling math questions like the one that Gauss's teacher posed to his class generations ago.

THE MONEY PROBLEM

Here is a classic puzzle that appears in many books. Look at the letters below.

(2) Can you figure out what numbers to insert for the letters to make the equation work? For each letter, only one number can be used. For example, anywhere that E appears, you must use the same number each time.

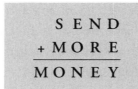

```
  S E N D
+ M O R E
---------
M O N E Y
```

THE NUMBERS GAME

A fun mathematics game that works well in years before 2000 is to take the numbers in a year and use those numbers to create equations that equal all of the numbers from 1 through 100. For example, using 1993, $1 = -1 - (9/9) + 3$, $2 = -1 \times (9/9) + 3$, $3 = [(1 + 9) - 9] \times 3$, and so on. A classic shorter version of this puzzle is the four 4s puzzle. For each number 1 through 10, write an equation using only four 4s but using any operations you choose. As an example for an equation for zero, you could use $4 - 4 + 4 - 4 = 0$. Another version of this puzzle is to substitute five 2s for your 4s.

AND ANOTHER THING  •••••••••••••

Sometimes math doesn't look like math because there are words, rather than numbers, involved. To figure out the words, though, you need to use your thinking skills, and math is always about thinking. So here's something like math fun, but involving words. Let's not call them word problems. Yuck! Call them *funzles*—fun puzzles.

Look at each word(s) picture and figure out a familiar phrase or word from the arrangement of the words. For example,

is "long division."

③ Can you figure out the funzles below?

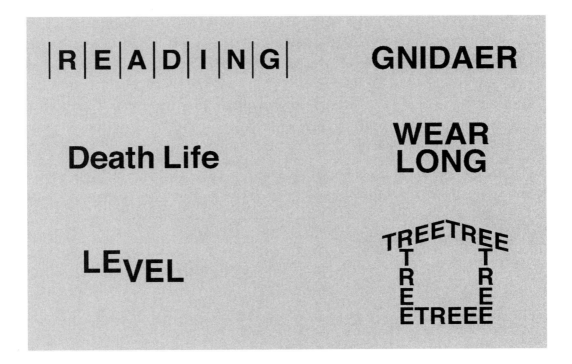

④ Can you make up five funzles of your own? Here are some phrases to get you started: "sleeping on the job" and "slowdown."

28 KEEPING COUNT ON A QUIPU

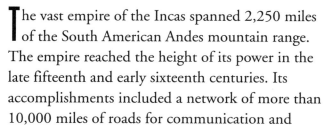

The vast empire of the Incas spanned 2,250 miles of the South American Andes mountain range. The empire reached the height of its power in the late fifteenth and early sixteenth centuries. Its accomplishments included a network of more than 10,000 miles of roads for communication and trade, use of terrace mountain farming techniques, and the building of great earthquake-resistant palaces and temples made by tight-fitting but irregularly shaped blocks.

Interestingly enough, despite these significant accomplishments, the Incas had no written number system. This did not prevent them from keeping elaborate records of the doings and events of every part of the empire.

Wise men called *armantus* kept numerical records of crop production, population changes, taxes, and the like on sets of knotted ropes called *quipus.* The type of information each system of knots carried was indicated through the color, thickness, and length of the ropes used. For example, a yellow rope might represent the maize crop. The Incans used a decimal system to record the numbers on their quipus with no knot representing zero. Historians are not sure how the decimal system was implemented on the ropes.

What number system would you use if you were making quipu records? Let's look at the number 264 to see how it would look in both the base 10 and base 2 number systems.

Base 10			Base 2								
2	6	4	256	128	64	32	16	8	4	2	1
2	0	0	1	0	0	0	0	1	0	0	0
	6	0									
		4									
2	6	4									

Base 2 totals (right side): 0, 0, 0, → 8, 0, 0, 0, 0, → 256, 264

(1) Based on the example above, how many ropes would you need to knot under base 10 to show 264?

(2) How many under base 2? Which seems more practical?

YOUR TURN

et's assume the Incas tied knots to represent numbers this way: ones—far right, tens—to the left of the ones, hundreds—to the left of the tens, and so on. This quipu represents the ears of maize grown in a farmer's field.

(3) If you use the system described above, how many ears of maize were grown according to the quipu record?

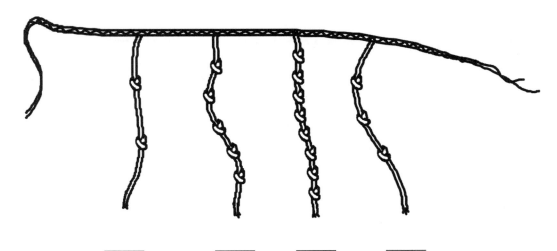

Make your own quipu by tying strings or colored yarn onto a cord. Tie on the population of a small Incan village in which 1,431 men, 2,123 women, and 1,601 children live. Make the cords different colors to represent each group of people represented.

(4) Try another recording challenge. Tie a record of knots in base 2, rather than base 10. Base 2 is the system upon which computers operate. Make a new quipu and tie on the number 45 in base 2. Hint: The number 10 in base 2 looks like this: 1010.

AND ANOTHER THING ....

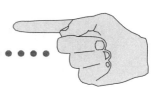

With limited writing tools available, ancient peoples were very creative about how they dealt with number information. For example, in ancient Mesopotamia, marketplace sellers and tax collectors had to figure out a way of adding and subtracting numbers without writing them down. This was because writing was cumbersome—using sticks to make indentations in soft clay tablets. For on-the-spot adding, they probably used a method of grooves and pebbles that was the forerunner of the abacus. Assuming they used a base 10 system, they would show the number 329 in three grooves.

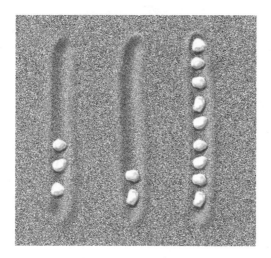

Make your own grooves in sand or dirt or fill a casserole dish with cornmeal or salt. You are a tax collector and need to total a farmer's wheat crops. You need to add 482 sheaves to 329. Put pebbles or beans in each groove to total 329 as shown above. Put beans totaling 482 in the appropriate grooves along with the 329 shown. Now move the beans around to show the total. Do this by deciding how many are too many beans in a particular groove, moving excess beans from ones to tens to hundreds, and so on. For example, if you have 11 beans in the ones groove, you move 1 bean to the tens groove, leave 1 bean in the ones groove and discard the other 9 beans. Then look at the tens groove and adjust it.

29 THE SYMMETRY OF THE TESSELLATED PLANE

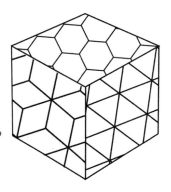

M. C. Escher, a Dutch artist, became famous for his tessellations of every imaginable creature and design. On a trip to Spain, he visited the Alhambra, a Moorish fortress housing palaces and gardens of the Moorish leaders. He became fascinated with the intricate mosaic designs created by Islamic artists. These artists were precluded by their religion from making decoration using animal or human images. As a result, they experimented with a wide array of geometric designs. Inspired by them, Escher spent hours copying the designs so he could study how they worked.

What is a tessellation? A tessellation is a pattern of shapes that completely covers a plane without any of the shapes overlapping or leaving open space between them. Certain polygons tessellate a plane without any gaps, as shown in these illustrations.

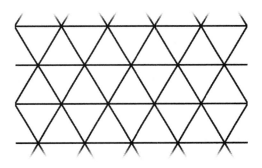
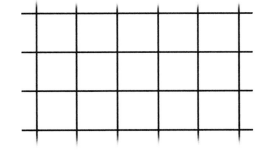

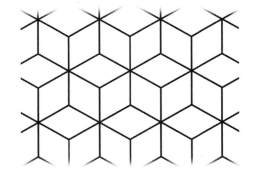
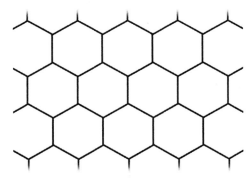

1. Can one of these figures look three-dimensional?

2. From the designs shown above, we know that certain figures with three, four, and six sides tessellate a plane. Can you figure out why?

A regular pentagon is a pentagon, the sides of which are all the same length.

(3) Can a regular pentagon tessellate a plane?

Try it with a piece of graph paper, a pencil, and a protractor. On a piece of graph paper, draw a pentagon. Now, try to figure out a way to connect other pentagons to it so that the paper is completely covered with pentagons with no gaps and no overlaps. Can you do it? Why? Why not?

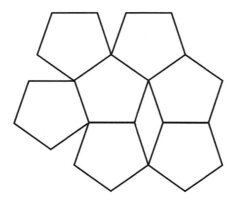

Let's see what the difference is between the figures on page 93 and the pentagon. Measure the angles of the triangle, hexagon, and square on page 93. How do the individual angles relate to the 360 degrees of a circle? Measure and total the angles at a vertex where each of these shapes join. How does the total of these angles relate to the 360 degrees of a circle? Now, what is the angle of the pentagon? Is there a different relationship to the 360 degrees of a circle?

Let's look more closely now at quadrilaterals. All the quadrilaterals we have looked at so far were parallelograms—meaning that their opposing sides were parallel to each other. Can you make another kind of quadrilateral that does not have parallel sides? Draw a scalene quadrilateral—one with no angles that are the same. See if you can tessellate a plane with the quadrilateral you drew.

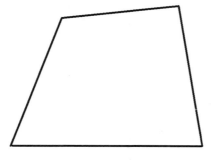

Now that you have an idea why some polygons tessellate a plane and others don't, try your hand at making an Escher-style tessellation. To do this, begin with a square. If you change one side of the square by deleting something, you must add that something back to the square on the opposing side so that the total area of the square remains the same and each piece will fit into the next.

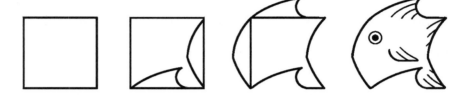

Can you apply some of the same principles to create a new tessellated shape from a triangle?

30 A NOT-SO-REGULAR GUY WHO PROVED THERE WERE FIVE REGULAR SOLIDS

Plato was a famous Greek philosopher who lived from 427 B.C. to 347 B.C. He was a very busy guy who was thinking all sorts of deep thoughts about what is real and what is not real. One of the things he thought about along the way was solid shapes. When he thought about shapes, he wasn't so worried about rocks, seeds, sticks, and pots. He knew these were solids, but he was interested in solids of certain shapes that are called *regular solids*.

What is a regular solid? It is a three-dimensional shape—a *polyhedron*—the faces of which are regular polygons of the same shape and size. What is a regular polygon? It is a two-dimensional shape, the edges or sides of which are all the same length and the angles of which are all the same such as equilateral triangles, squares, and regular pentagons. What Plato figured out is that there are exactly five regular solids that can be made with these regular polygons. Today we call them the *platonic solids*.

What are the five platonic solids? They are the tetrahedron (four faces), the hexahedron or cube (six faces), the octahedron (eight faces), the dodecahedron (twelve faces), and the icosahedron (twenty faces).

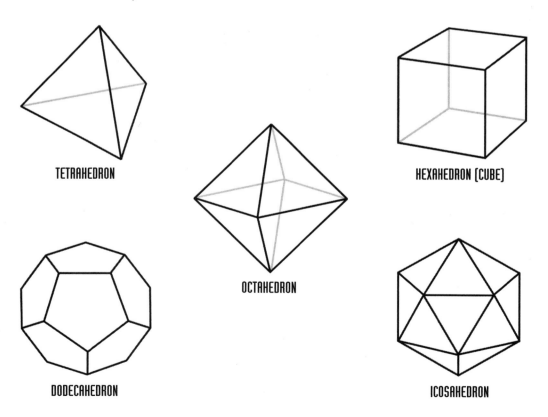

TETRAHEDRON

HEXAHEDRON (CUBE)

OCTAHEDRON

DODECAHEDRON

ICOSAHEDRON

Here are the patterns for making a dodecahedron and an isocahedron.

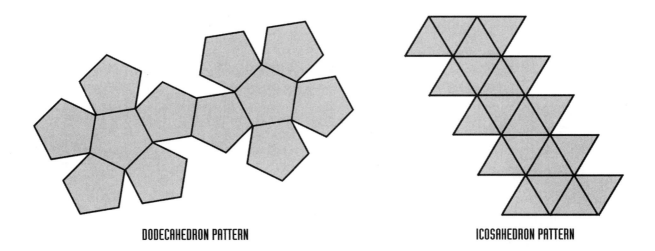

DODECAHEDRON PATTERN   ICOSAHEDRON PATTERN

YOUR TURN ·

Making patterns to create the regular solids can be fun and challenging.

(1) Can you design patterns for making a tetrahedron, hexahedron, and octahedron?

Use the regular polygon shape that makes up the solid to create a pattern that you can trace onto heavy paper or cardboard to make a three-dimensional solid.

To create your patterns, you need graph paper, a pencil, and a ruler. Let's begin with the tetrahedron. Look at the picture of this polyhedron's shape. How would you lay out the four triangles on paper so that they could be cut out in one piece and folded to form this shape?

Begin with the base. Draw it on your graph paper. Then figure out how the other three faces relate to that base face. Once you have created the pattern for the four-faced solid, try to create patterns for the other platonic solids.

As you look at your platonic solids, can you come up with any numerical pattern for the way the number of faces and corners (or vertices) relate to the number of edges in these solids? To get you started, consider the tetrahedron and the cube:

	Number of Faces	Number of Corners	Number of Edges
Tetrahedron	4	4	6
Cube	6	8	12

(2) What is always true about the relationship between the sum of the total number of faces and corners to the total number of edges? Make up a formula that describes the relationship.

Visualize a six-faced solid made with equilateral triangles.

③ Can you figure out how to make a pattern for such a polyhedron? Look at the tetrahedron to get you started.

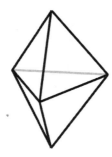

Go on a platonic solid search. What shape are salt crystals? What shape are sugar crystals? Check out the Egyptian pyramids.

31 | WHAT PYTHAGORAS SAID

The Greek philosopher and mathematician Pythagoras (580 B.C.–500 B.C.) gave the world one of the most famous *theorems* (or rules) of mathematics. He was puzzled by the Egyptian right triangle.

What is the Egyptian right triangle? In ancient Egypt, architects and builders needed to use right, or 90-degree, angles in building structures such as pyramids and temples. The tool they used to find the proper angle was made of three knotted ropes. Rope stretchers pulled three ropes to form a right-angle triangle—one rope knotted to show three units, one knotted to show four units, and one knotted to show five units. The angle made by the intersection of the two short sides of the triangle was always a right angle.

In looking at the Egyptian right-angle triangle, Pythagoras found that the squares of each of the short sides totaled the square of the long side.

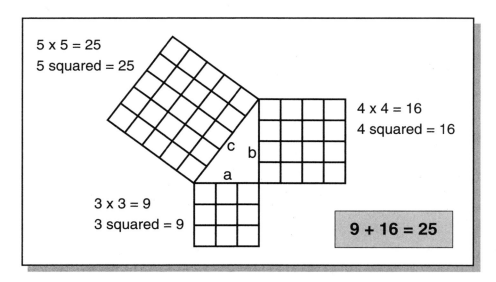

5 x 5 = 25
5 squared = 25

4 x 4 = 16
4 squared = 16

3 x 3 = 9
3 squared = 9

9 + 16 = 25

Pythagoras wondered if his formula

$$a^2 + b^2 = c^2$$

worked for all right-angle triangles. Pythagoras proved that it did, and you can too.

YOUR TURN ....

Take a look at this geometric design, which is a series of squares divided into triangles. Can you use it to prove that Pythagoras's theorem works for right-angle triangles with two equal sides? Can you do Pythagoras's proof with these shapes, which might be found in any common floor tile? Make the design in the illustration below (at right) on graph paper and cut out the triangles to prove Pythagoras right.

$$a^2 + b^2 = c^2$$

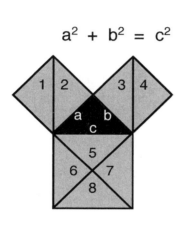

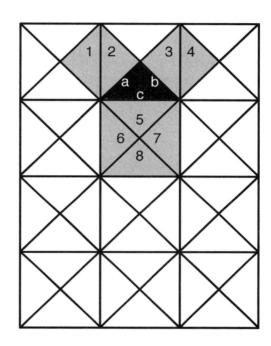

All the triangles are right-angle triangles having two equal sides. This is because diagonals of a square are perpendicular to each other or at 90-degree angles. Can you see that each of the triangles numbered 1 through 8 has the same area? This is because all of these triangles are the same size.

Now look closely at triangle abc. When triangles 1 and 2 are taken together, they are the square of side a. Similarly, triangles 3 and 4 are the square of side b and triangles 5, 6, 7, and 8 are the square of side c. Do triangles 1, 2, 3, and 4 have the same area as triangles 5, 6, 7 and 8? In other words, when put together, are triangles 1, 2, 3 and 4 equal to triangles 5, 6, 7, and 8? If they are, you have proved the Pythagorean theorem for right triangles with two equal sides.

AND ANOTHER THING

Okay, so Pythagoras was right about the right triangle with two equal sides, but does the theorem really apply to all right triangles? You can prove that it does.

▶ On a piece of graph paper, make a square, each side of which is 2 inches in length.

▶ Draw two straight lines that intersect at right angles so that, within the framing square, you have a large square (square B), a small square (square A), and two equal rectangles (rectangles D and E).

▶ Now divide your two rectangles into triangles by drawing one diagonal line through each. Label them as shown in the illustration below.

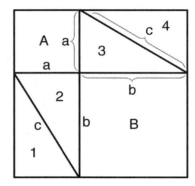

Square A shows the square of side a on triangles 2 and 3. Square B shows you the square of side b on triangles 2 and 3. How can you find the square of side c? It should equal the combined area of squares A and B if Pythagoras was right. Let's find out.

On your graph paper, draw a second framing square the same size as the first one you drew. Cut out triangles 1, 2, 3, and 4, and place them in the corners of the framing square so that they form an interior square (square C).

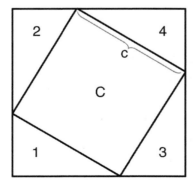

Doesn't this square have the same area as the area of squares A and B taken together? Voilà! You have proven the Pythagorean theorem. Show that Pythagoras was right for any right triangle by doing this activity again, this time placing your intersection point at a different place within the framing square.

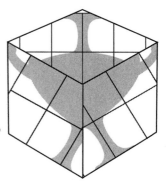

32 THE GOLDEN RECTANGLE

The *golden rectangle* is a rectangle of a certain shape that, through history, has been considered very pleasing to the eye. While there is no evidence that the ancient Greeks thought about this rectangle, they used it in building the Parthenon, a temple that stands on the Acropolis in Athens, Greece. The Parthenon has long been lauded as one of the most beautiful structures in the world. Built in the fifth century B.C., the proportion, or overall size and shape of the structure of this building, is based upon the root 5 rectangle, now known as the *golden rectangle*.

The golden rectangle is based upon the *golden mean*, which has also been called the *golden section* or *golden ratio*. The beauty of the rectangle and the related ratio is seen again and again in architecture, art, nature and even the proportions of human anatomy.

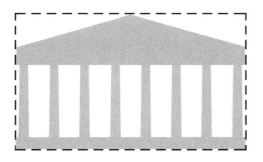

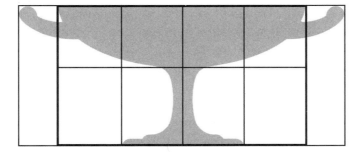

Many artists, such as Leonardo da Vinci and Piet Mondrian, have used it in their paintings. It is said that if you draw a rectangle around the face of da Vinci's "Mona Lisa," the rectangle is golden.

Okay, but what is a golden rectangle? It is a rectangle like any other, but it has specific proportions. It's not too thin. It's not too long. It's just right. The sides of this "just right" rectangle are made when its length is a little more than 1½ times its width-that is, the ratio of width to length is approximately 1 :: 1.618.

Let's see how to make a rectangle with this proportion. If you could make a line segment that was divided into the golden section, you would have the length and width you need to make the rectangle. A golden section looks like this.

In this line segment, the length of BC is to the length of line segment AB as the length of AB is to the length of AC. Mathematically, you write it in this way:

/BC/ / /AB/ = /AB/ / /AC/

Said another way, the smaller part, $\overline{BC}$, is to the larger part, $\overline{AB}$, as the larger part, $\overline{AB}$, is to the whole, $\overline{AC}$.

YOUR TURN

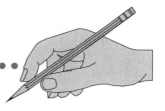

How can you make a golden rectangle without measuring? You can make one in this way:

► Make any square ABCD.

► Bissect the square with line segment $\overline{JK}$.

► Make an arc using a compass. The center of the arc is K and the radius is $\overline{KB}$.

► Make the arc into a semicircle ending at G.

► Draw the arc downward through line $\overline{DC}$ so that you can extend $\overline{DC}$ to point E to make $\overline{DE}$.

► Draw a ray perpendicular to $\overline{DE}$ so that the extension of $\overline{AB}$ intersects it at F.

► Now you have golden rectangle AFED and golden section $\overline{CE}$ is to $\overline{DC}$ as $\overline{DC}$ is to $\overline{DE}$.

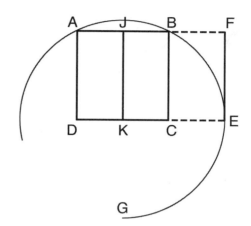

(1) Look at this picture of the Parthenon. Can you figure out how the Parthenon uses the golden rectangle within its overall structure? Specifically, can you find two large golden rectangles and four small ones?

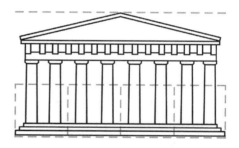

AND ANOTHER THING

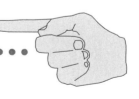

The golden ratio and the golden rectangle are seen in the paintings of many great artists. Something about the shape is aesthetically pleasing to human eyes. Find some of the later artworks of twentieth-century Dutch painter Piet Mondrian. See if you can find golden rectangles and golden ratios in the structure of his geometric paintings. Create your own geometric designs. In one, use the golden ratio and golden rectangle. Try to avoid these proportions in another. Which one appeals more to your eye?

33 GOING AROUND IN CIRCLES

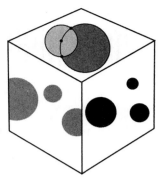

Since the time of the ancients, geometric shapes and their relationships have fascinated math thinkers and philosophers. When they tried to figure out how geometric shapes related to one another, they were limited in their tools. They had no computers. They had no calculators. They didn't even have rulers with measuring lines. When they wanted to solve mathematical problems, they had to do so with only a straightedge (sort of like a ruler but with no measuring lines) and a compass-like device (a stick with a piece of string). Of course, they also had their brains. They thought about these problems a lot.

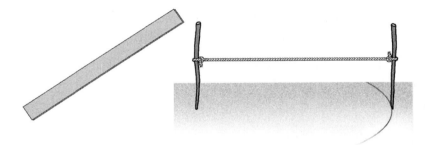

The fact that the ancients had limited tools didn't stop them from thinking up very tough problems. In fact, there are many famous problems that the thinkers of antiquity struggled with, which continued to puzzle great thinkers for hundreds of years. Surprisingly, or perhaps not so surprisingly, mathematicians found that there were no easy answers to some of these problems.

One of the most famous of these problems has been called "squaring the circle." In other words, how can you make a square with the same area as any given circle?

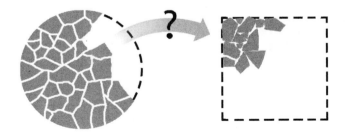

In an effort to solve this and other math problems, great math thinkers through history made many interesting discoveries—even though many of these discoveries did not help solve the problems they were investigating. Often, the problems the ancients thought of had to be solved using tools they did not have at their disposal, or they needed higher math skills developed over time such as a student of advanced math analysis might possess today.

YOUR TURN ·

ven though you might not be able to solve the problems that stumped Nicomedes, Archimedes, and other great Greek and Roman thinkers, you can use a straightedge and compass to have fun solving some other problems. Remember, the ancients did not have rulers; they had straightedges.

(1) Here is a circle problem that you might enjoy. If you know only the circumference of a circle such as the circle below, how do you find the center point (or origin) using only a compass and straightedge?

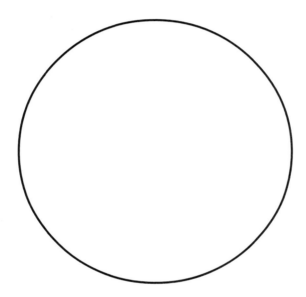

Hint: Begin by finding a point on the circumference of the circle and drawing another circle from that point that intersects the first circle in two places.

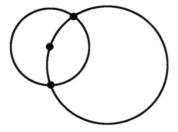

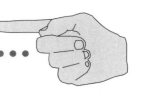

Another great ancient thinker named Apollonius of Perga had his own problem with circles. His problem came to be known as—can you believe it—Apollonius's problem: For any three circles, find a fourth circle that is tangent to all three. An example is the set of circles below. On the left are the three circles to be joined by the fourth. On the right is the fourth circle tangent to all three.

(2)   Can you find one or more fourth circles that solves Apollonius's problem for the set of three circles below?

34 A FIBONACCI SEARCH

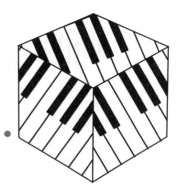

A famous mathematician named Leonardo of Pisa (Pisa, Italy, was his birthplace) was born in about 1175 and lived until about 1240. When he wrote his books and treatises on math, he used the name Fibonacci and he is best known by this name.

As a young person, Fibonacci traveled with his father, who was a customs agent. In these travels, he became familiar with Hindu-Arabic numerals, which used both a zero and place value. He thought this number system was far more practical than the Roman numeral system that was then still in use in Italy. What do you think? Is it easier to figure out the sum of 36 + 14 or the sum of XXXVI + XIV? Fibonacci wrote about the Hindu-Arabic number system and helped influence its adoption throughout Italy and, ultimately, the Western world.

(1) Fibonacci also like to think up interesting problems that involved math. Through one of these problems, he discovered a sequence of numbers that is very interesting. An example of the *Fibonacci sequence* goes like this: 1, 1, 2, 3, 5. Can you figure out the next number in the sequence?

The Fibonnaci sequence is the series of numbers made by adding the two prior numbers to get the next number, beginning with 0 + 1. Now can you figure out the next number in the Fibonnaci sequence above?

(2) Here is another sequence of numbers made the same way: 4, 4, 8, 12, 20, 32. What is the next number in this sequence?

YOUR TURN

The Fibonacci numbers occur surprisingly often in the world around us. Look for Fibonacci numbers on a piano. Find middle C. To the right of middle C, how many black keys are grouped together?

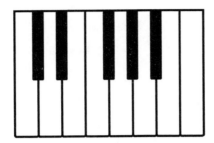

How many black keys are there in the next group? Look at the black and white key pattern and see how many Fibonnaci numbers you can find.

Do a plant search for Fibonacci numbers on living plants. Look at how the leaves grow. Do they spiral around the stem? Moving upward vertically on the stem, how many leaves are there before a leaf appears on the stem directly above the first leaf you counted? In what number pattern do the twigs grow from the branches of a tree or shrub? How many petals are on the flowers? Look at several plants. How many seem to grow in Fibonacci numbers?

Take a look at an artichoke, a pineapple, or a pinecone. Look for the spiraling pattern of the petals, scales, or bracts. Do the rows of these coverings occur in Fibonacci numbers?

Look at the bones of your body. Can you find Fibonacci number relationships in the bones of your hands? Your feet?

AND ANOTHER THING ....

A little problem about rabbits started Fibonacci on his number search. Here is how the problem went:

(3) If you began the year with a pair of rabbits, how many pairs of rabbits would there be after a year if you assume that every month each pair of rabbits produces one new pair of rabbits and each of these rabbits bear young two months after birth?

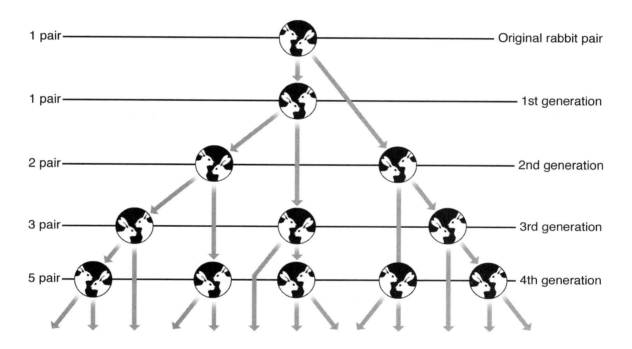

You can begin by drawing a diagram to find the answer, but soon your picture will become unmanageable. Can you find the answer by using what you know about the Fibonacci number sequence?

35 FROM NUMBER PLACE TO SUDOKU— EVERY NUMBER HAS ITS PLACE

Games and puzzles come and go in popularity. Some games that have been around for years suddenly become all the rage. One such game is a little like a magic square, but then again, not like it at all. It is known as *Sudoku*.

Dell Magazines originated the first Number Place puzzle more than 25 years ago. The puzzle later caught on in Japan with the name *Sudoku*, and now puzzle lovers worldwide play *Sudoku*. *Su* means "number" in Japanese. *Doku* loosely means "bachelor" or "single." *Sudoku* can be translated loosely as "single number."

A Sudoku puzzle contains nine 3 x 3 squares inside a 9 x 9 square and looks like this:

The small squares are called *cells*. The 3 x 3 squares are called *squares*. A horizontal row of nine cells is called a *row*. A vertical line of nine cells is called a *column*. The entire grid is called the *puzzle*.

The rules are simple. Place the digits 1 through 9 in the cells so that each digit occurs only once in each square, row, and column. Here is a finished puzzle:

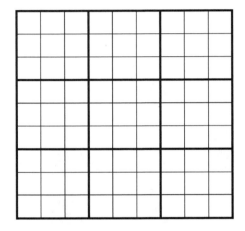

Other versions of this puzzle have appeared through the years. Wordoku is a twist on Sudoku, using letters instead of numbers. You can also substitute shapes for numbers.

YOUR TURN

Now try a few Sudoku puzzles on your own.

Puzzle 1

	3			6				7
	4		2	5	3	9		6
	5	6	7		2			
9	2	5			8	7		1
	6						2	
4		7	9			6	3	5
		2			9	4	6	
6	9	8	3		7		5	
5			8				7	

Puzzle 2

5	9		7	8				6
		7	2	1				5
2				9	8	7		
4		9		1		3		
	2		9		7		4	
	1		3			5		9
	7	2	1					4
1				9	2	3		
3				7	5		2	1

For more puzzles, look on the Internet, in newspapers and magazines, and in bookstores. Sudoku is everywhere!

AND ANOTHER THING

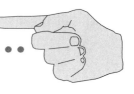

Do you use the same strategies to solve the puzzle when you use shapes as when you use numbers? Try this shape Sudoku puzzle.

✚			♥			★	☽	
♥	●	♣		✚			■	
		◆		■		●		
✚		★		●		■		◆
	♣		▲		◆		♥	
■		♥		★		▲		●
	✚		■		♣			
	♥		☽			●	◆	♣
♣	☽			♥				

ANSWER KEY ........................

1 FACTORING, GEOMETRIC SHAPES, AND PRIME NUMBERS

1. The factors for 5 are 1 and 5. The factors for 12 are 1, 12, 2, 6, 3, and 4.

2. Yes, 5 is a prime number because it has only two factors, itself and 1, but 12 has six factors, so it is not a prime number.

2 THE SHAPE OF NUMBERS—IT'S GREEK TO ME

1. When you add consecutive triangular numbers, you get a square number.

2. When you add consecutive odd numbers, you get square numbers:

   $1 = 1$
   $1 + 3 = 4$
   $1 + 3 + 5 = 9$
   $1 + 3 + 5 + 7 = 16$ and so on.

3. A cubic number is any number multiplied by itself twice. It is shown with exponents as in the case of 2 x 2 x 2, or 2 to the third power (2^3), or 8.

3 ERATOSTHENE'S PRIME NUMBER SIEVE

1. The prime numbers between 1 and 100 are 2, 3, 5, 7, 11, 13, 17, 19, 23, 29, 31, 37, 41, 43, 47, 53, 59, 61, 67, 71, 73, 79, 83, 89, and 97.

4 A HEXAGON HERE, A HEXAGON THERE

1. When bubbles join in foam, they coalesce into hexagonal shapes.

5 BONING UP ON RATIOS

Ratio answers are reduced to their smallest number.

1. Ratio of upper arm bones to lower arm bones, 1:2

2. Ratio of finger phalanges to toe phalanges, 1:1

3. Ratio of ankle bones to toe bones, 1:2

4. Ratio of nasal bones to total skull bones, 1:6

5. Ratio of ankle and instep bones to toe bones, 6:7

6. Ratio of lumbar vertebrae to sacrum vertebrae, 1:1

7. Ratio of adult bones to bones in a baby, 206:350

8. Rounded (nearest 50) ratio of adult bones to bones in a baby, 200:350

9. Quick useful ratio of adult to baby bones, 4:7

6 IS SYMMETRY JUST SYMMETRY?

1. Equilateral triangle: rotate 120 degrees; square: rotate 90 degrees; hexagon: rotate 60 degrees

7 THE MAGIC LINE—A MATTER OF SYMMETRY?

Magic lines of magic squares.

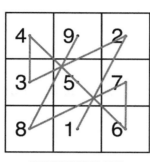

ANCIENT TIBETAN SEAL

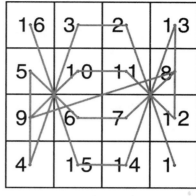

DURER'S MAGIC SQUARE

1. Rotational symmetry; 180 degrees

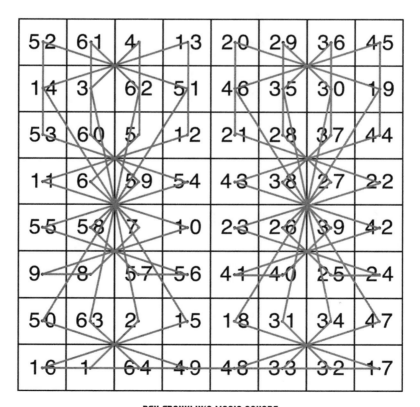

BEN FRANKLIN'S MAGIC SQUARE

2. Reflective symmetry

9 STRAIGHT TO THE ARC

1. The arc of a circle is formed using the 90-degree angle.
2. Yes. Square: reflective and rotational symmetry
3. Equilateral triangle: reflective and rotational symmetry
4. Eight-point star: reflective and rotational symmetry; broken cross design: rotational symmetry

10 ONE SIDE OR TWO——THE PAPER BAND MEETS AUGUST MÖBIUS

1. Two sides
2. One side
3. Two bands with two sides each
4. Here is how you cut the Klein bottle in half to make a Möbius strip:

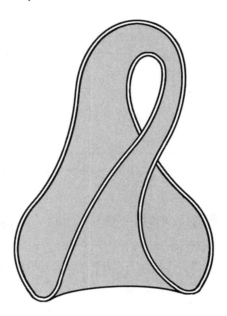

14 THE GEOMETRIC FOLD——SHAPES WITHIN SHAPES

1. To make a pentagon with a 1-by-10-inch paper strip, fold it into a simple knot.

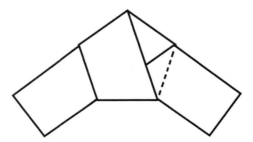

2. Fold an ellipse by placing a dot on the circle. Do not place the dot in the center. Work your way around the circle, folding the circle's edges to the dot until you have made your way around the circle.

16 NETWORKS NOT ON TELEVISION

1. The traversible networks of the group shown are the top left and bottom left figures.

2. From left to right, the first and third figures are traversible and the second and fourth figures are not.

3. Turn the Konigsberg bridge path into a network. Each area of land becomes a point or vertex. Draw lines from each vertex. The resulting network has four odd vertices and is not traversible.

18 PUZZLES AND GAMES—THE TOOTHPICK WAY

1. A subtraction puzzle

2. Lose that square

3. A triangular puzzle

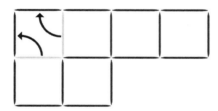

4. Triangles and squares

5. Move one toothpick to solve the equations.

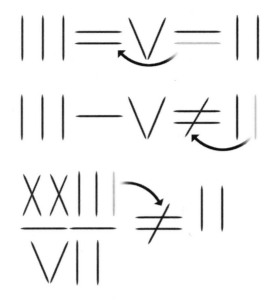

6. Move four toothpicks to create three triangles.

7. Move four toothpicks of the spiral to make three squares.

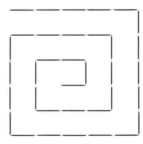

8. Move three toothpicks to create three squares.

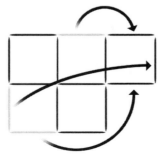

9. Create four more squares with twelve toothpicks. Remove four toothpicks to leave nine squares.

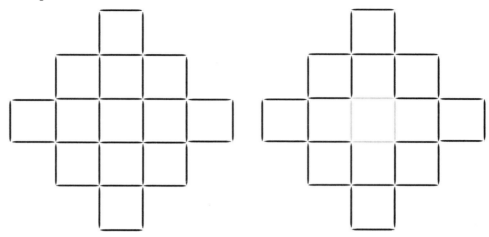

10. Turn the triangle upside down by moving three coins.

11. Move one coin to form two rows of four coins.

12. Rearrange the coins to form a square with five coins on each side.

19 UNPUZZLING THE TANGRAM

1. Use these directions for making a tangram puzzle.

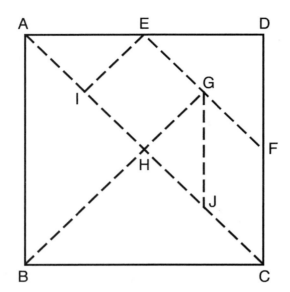

a. Find the midpoint between A and D and label it E. Then find the midpoint between D and C and label it F.

b. Find the midpoint between E and F, label it G, and connect it with B.

c. Connect A and C. Find the midpoint of $\overline{AC}$, and label it H. (This is where $\overline{BG}$ intersects $\overline{AC}$.)

d. Find the midpoint between A and H. Label that midpoint I, and connect it with E.

e. Find the midpoint between H and C, label it J, and connect G with J.

2. Solutions to six polygonal shapes:

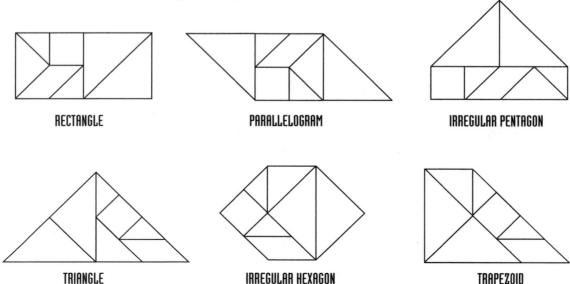

RECTANGLE   PARALLELOGRAM   IRREGULAR PENTAGON

TRIANGLE   IRREGULAR HEXAGON   TRAPEZOID

3. Solution to tangram puzzle:

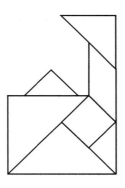

COMPUTER DESK

4. Solutions to tangram animal puzzles:

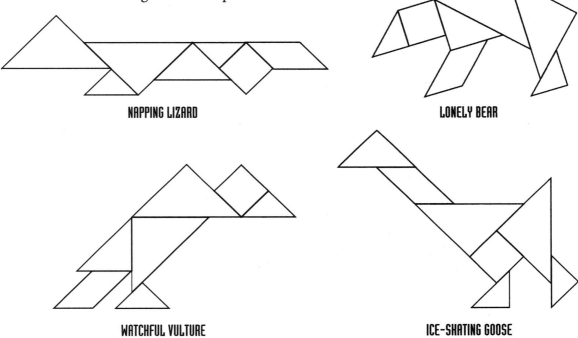

NAPPING LIZARD

LONELY BEAR

WATCHFUL VULTURE

ICE-SKATING GOOSE

5. Solutions to three more puzzles:

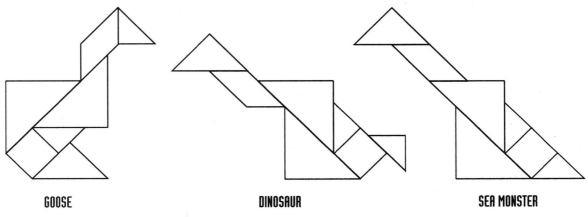

GOOSE

DINOSAUR

SEA MONSTER

19 UNPUZZLING THE TANGRAM [continued]

6. Sample eight-piece rectangular puzzle

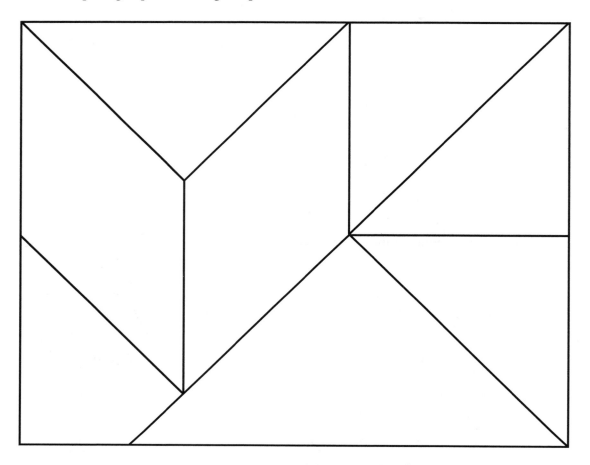

20 HOW THE EGYPTIANS MULTIPLIED

1. $\frac{1}{2} + \frac{1}{4} + \frac{1}{16}$

21 MAGIC SQUARE IN THE MAKING

1.

17	24	1	8	15
23	5	7	14	16
4	6	13	20	22
10	12	19	21	3
11	19	25	2	9

2. The stairstep method works only on odd magic squares—that is, squares that are three-by-three, five-by-five, and so on.

3.

6	2	5
3	1	4
8	7	9

23 WHAT'S KNOT TO LIKE?

1. The left and right trefoil knots are not the same knot.

2. These are the same knot. The knot has four crossings.

3. These are not the same knot. The knot on the left is a granny knot. The knot on the right is a reef knot.

24 JORDAN CURVE? CAN YOU THROW ONE?

1. In the spiral-shaped Jordan curve, point B is inside the curve and point A is outside the curve.

2. In the odd-shaped Jordan curve, point A is inside the curve and point B is outside the curve.

3. The method works on polygons. In the polygonal Jordan curve, point A is inside the curve and point B is outside the curve.

25 AN A-MAZE-ING THING

1. Sample maze designs for the networks

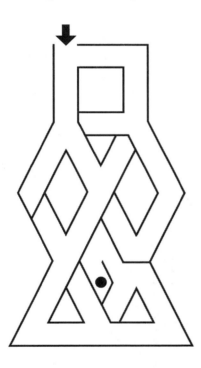
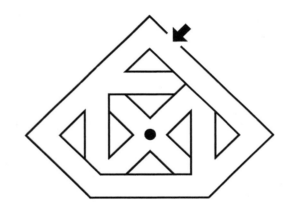

2. Maze solution (Both designs can make this maze.)

3. The Knossos maze is not a Jordan curve, because it contains an intersecting line.

26 IS IT PROBABLE?

1. When you toss one coin, there are two possible results. When you throw a single die, there are six possible results.

2. Completed Pascal triangle

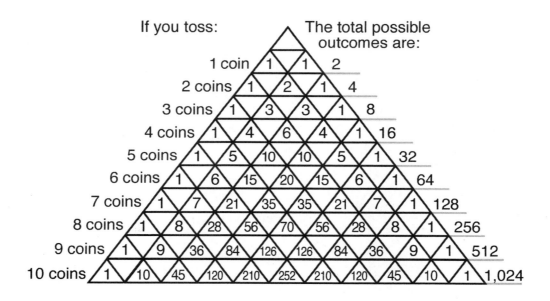

3. If you add the numbers on the diagonal lines drawn through Pascal's triangle, you get a Fibonacci number sequence.

4. Some things to look for in Pascal's triangle: 1) Each horizontal row is a palindrome, and 2) the total possible outcomes double with each coin you add to the toss.

27 CAN YOU GAUSS THE ANSWER?

1. To add 1 through 100 quickly, pair each number 1 + 100, 2 + 99 and so on to 50 + 51. Each pair totals 101 and there are 50 pairs. The answer is 5,050.

2. SEND   9567
    + MORE   1085
    MONEY   10,652

3. Funzle solutions:

   reading between the lines

   life after death

   split level

   reading backward

   long underwear

   tree house

4. sleeping on the job: sleeping

                          the job

   slowdown: S

                          L

                          O

                          W

Consider these additional funzle ideas:

▶ Write the word *game* using nothing but stars = "all-star game."

▶ Write the word *teen* using nothing but numeral 6 = "sixteen."

▶ Write the word *cycle* twice = "bicycle."

28 KEEPING COUNT ON A QUIPU

1. 3

2. 9

3. 2, 583 ears of maize

4. Quipu knots for 45 in base 2 looks like this:

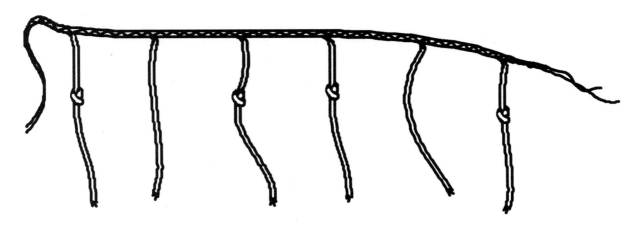

29 THE SYMMETRY OF THE TESSELLATED PLANE

1. When shaded, the diamond tessellation looks three-dimensional.

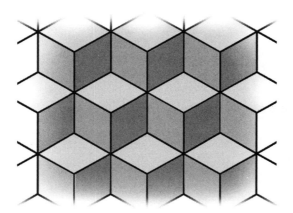

2. The angles at the vertex of polygons that will tessellate a plane total 360 degrees.

3. Each angle of the equilateral triangle, regular hexagon, and square can be divided into 360 degrees with no remainder. When the shapes are joined at a vertex, the angles at the vertex total 360 degrees. The angles at the vertex of the pictured pentagons do not total 360 degrees. Accordingly, a regular pentagon cannot tessellate a plane.

30 A NOT-SO-REGULAR GUY WHO PROVED THERE WERE FIVE REGULAR SOLIDS

1. Patterns for regular solids

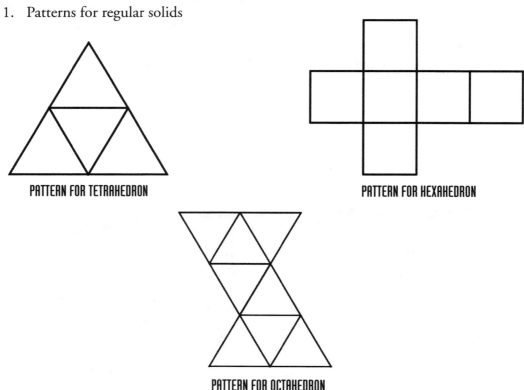

PATTERN FOR TETRAHEDRON

PATTERN FOR HEXAHEDRON

PATTERN FOR OCTAHEDRON

2. A formula for the faces/corners/edges question is as follows:

number of faces + number of corners = number of edges minus 2

The total number of edges is two less than the combined total number of faces and corners.

3. Pattern for six-faced solid made from equilateral triangles

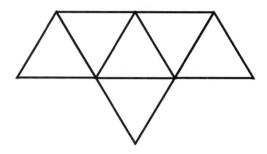

32 THE GOLDEN RECTANGLE

1. How the Parthenon uses the golden rectangle

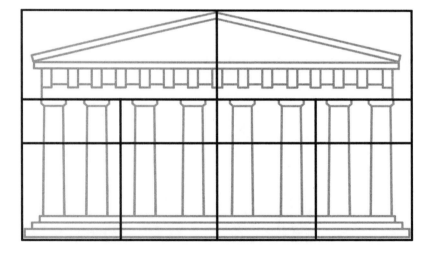

1. For this one, you need only a compass to take the following steps.

    i. Draw a circle. Make a circle starting at any point A on the circumference of the circle. Identify the two points where this circle intersects the first circle as points B and C.

    ii. Use points B and C as the center to make arcs with a radius of $\overline{AB}$ and $\overline{AC}$.

    iii. Mark as D the point at which these arcs intersect inside the original circle. Make a circle with the radius $\overline{DA}$.

    iv. Draw line segment $\overline{DA}$ and continue it until it intersects the side of the circle opposite of A. This intersection point E will be the center (or origin) of the original circle.

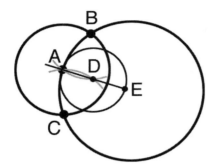

2. Sample solutions to Apollonius circle problem:

How many more can you find?

34 A FIBONACCI SEARCH

1. 8
2. 52
3. The answer to the rabbit problem is 377 rabbits.

35 FROM NUMBER PLACE TO SUDOKU—EVERY NUMBER HAS ITS PLACE

The answers to the Sudoku puzzles:

2	3	9	4	1	6	5	8	7
7	4	1	2	8	5	3	9	6
8	5	6	7	9	3	2	1	4
9	2	5	6	3	8	7	4	1
1	6	3	5	7	4	8	2	9
4	8	7	9	2	1	6	3	5
3	7	2	1	5	9	4	6	8
6	9	8	3	4	7	1	5	2
5	1	4	8	6	2	9	7	3

5	9	4	7	8	3	2	1	6
8	3	7	2	1	6	4	9	5
2	6	1	5	4	9	8	7	3
4	5	9	8	6	1	7	3	2
6	2	3	9	5	7	1	4	8
7	1	8	3	2	4	5	6	9
9	7	2	1	3	8	6	5	4
1	4	5	6	9	2	3	8	7
3	8	6	4	7	5	9	2	1

The answer to the Sudoku shapes puzzle:

GLOSSARY

arc: part of the circumference of a circle

concave: inward curve

congruence: a condition in which objects are the same size and shape

convex: outward curve

factor: number that can be divided into a whole number with no remainder

geodesic dome: domed structure made of bars or segments that form interlocking polygons

geometry: the study of surfaces, points, lines, and curves

golden rectangle: a rectangle that, if divided by one line into a square and a resulting rectangle, the resulting rectangle will be similar to the original rectangle

integer: a positive or negative number, including zero, that does not have a decimal or fraction

Jordan Curve: a simple closed curve, a closed curve being one that does not intersect itself

magic square: a square arrangement of numbers into rows and columns such that the sum of each row, column, or diagonal is the same

network: a path that is made with a series of lines and vertices

palindrome: a phrase, word, or number that reads the same forward or backward

parabola: a curve that is made by the intersection of a plane parallel to an element of a cone

platonic solid: one of five solids, the faces of which are regular polygons

polygon: closed figure formed by three or more line segments; a regular polygon is one in which all of the line segments are the same and all the angles are the same

polyhedron: a three-dimensional shape that has faces or sides made of regular polygons of the same size and shape

prime number: a whole number greater than 1 that has only itself and 1 as factors

probability: the chances of a specific result occurring

Pythagorean Theorem:	theorem stating that in a right triangle, $a^2 + b^2 = c^2$, where c is the hypotenuse, the longest side of the triangle opposite the right angle
similarity:	in figures, when their corresponding angles are equal and their corresponding sides are proportional
square root:	the square root for a number is that number whose square is the number
symmetry:	a condition in which a form corresponds exactly on opposite sides of a dividing line
tangent:	making contact at a single point or along a line, without intersecting
tangram:	a geometric Chinese puzzle in which seven geometric shapes fit within a square and can be used to make other designs
tessellation:	a repeated geometric design that covers a plane with no overlaps or gaps
theorem:	a statement for which there is a proof
topology:	the study of geometric shapes that are not changed by stretching or bending
vertex:	a point at which the sides of an angle intersect or a point at which three or more sides intersect in a polyhedron

BIBLIOGRAPHY

Britton, Jill and Walter Britton. *Teaching Tessellating Art*. Palo Alto, CA: Dale Seymour Publications, 1992.

Courant, Richard and Robbins, Herbert. *What Is Mathematics?* New York: Oxford University Press, 1996.

Fixx, James. *Games for the Super-intelligent*. New York: Warner Books, 1991.

Gardner, Martin. *The Colossal Book of Mathematics: Classic Puzzles, Paradoxes and Problems*. New York: W. W. Norton & Company, 2001.

Godwin, Edward. *Number Mania: Math Puzzles for Smart Kids*. New York: Sterling Publishing, 2002.

Know Your Body: The Atlas of Anatomy. Berkeley, CA: Ulysses Press, 1999.

Lee, Martin. *40 Fabulous Math Mysteries Kids Can't Resist*. New York: Scholastic, 2001.

The New Encyclopedia Britannica. Chicago: Encyclopedia Britannica, 2003.

Pappas, Theoni. *The Joy of Mathematics*. San Carlos, CA: Wide World Publishing, 1989.

Pappas, Theoni. *More Joy of Mathematics*. San Carlos, CA: Wide World Publishing, 1991.

Savadori, Mario with Joseph Wright. *Math Games for Middle School: Challenges and Skill-Builders for Students at Every Level*. Chicago: Chicago Review Press, 1998.

Slavin, Steve. *All the Math You'll Ever Need: A Self-Teaching Guide*. New York: John Wiley & Sons, 1999.

INDEX